T0123045

# Spark from the Deep

ANIMALS, HISTORY, CULTURE
Harriet Ritvo, *Series Editor*

# SPARK FROM THE DEEP

*How Shocking Experiments with Strongly Electric Fish Powered Scientific Discovery*

WILLIAM J. TURKEL

The Johns Hopkins University Press

*Baltimore*

Printed in the United States of America on acid-free paper
9 8 7 6 5 4 3 2 1

The Johns Hopkins University Press
2715 North Charles Street
Baltimore, Maryland 21218-4363
www.press.jhu.edu

Turkel, William J. (William Joseph), 1967–
Spark from the deep : how shocking experiments with strongly electric fish powered scientific discovery / by William J. Turkel.
pages cm — (Animals, history, culture)
Includes bibliographical references and index.
ISBN 978-1-4214-0981-8 (hardcover : alk. paper) — ISBN 978-1-4214-0994-8 (electronic) — ISBN 1-4214-0981-X (hardcover : alk. paper) — ISBN 1-4214-0994-1 (electronic)
1. Electric fishes. 2. Electricity—Experiments—History. 3. Discoveries in sciences—History. I. Title.
QL639.1.T87 2013
597—dc23 2012038726

A catalog record for this book is available from the British Library.

*Special discounts are available for bulk purchases of this book. For more information, please contact Special Sales at 410-516-6936 or specialsales@press.jhu.edu.*

The Johns Hopkins University Press uses environmentally friendly book materials, including recycled text paper that is composed of at least 30 percent post-consumer waste, whenever possible.

*For Juliet,*

*and in memory of Mark Edwin Sitton, 1974–1998*

# Contents

# Acknowledgments

The impetus for this book came from a graduate seminar that I took in the Brain and Cognitive Sciences department at the Massachusetts Institute of Technology in 1997. We were watching a film demonstrating Hubel and Wiesel's orientation-selectivity experiments. On the screen, a greenish-white bar turns on and off and rotates against a dark background. As it does so, a static noise becomes more or less pronounced. The film shows the stimulus being presented to an anesthetized cat while microelectrodes in its visual cortex audify its neuronal response. I was struck by the contrast between the banality of the images and sounds in the film and the wonder and horror of the activities that were required to create it in the first place. I decided I would rather be studying science as an activity than doing experiments; this book is my attempt to explain the origins and wider significance of electrophysiological experimentation.

As with my first book, this one would not exist if Harriet Ritvo had not read a messy collection of desultory notes and then captured what the whole thing could become in one brilliant, incisive phrase. She remains the best friend and mentor anyone could ever hope to have. My thinking was also shaped during many conversations with other close friends: Edward Jones-Imhotep on the history of electronics and the methodology of science, technology, and society; Rob MacDougall on big history and the history of technology; Tim Hitchcock on computational methods and new ways of doing history; Kevin Kee on "the method"; and Devon Elliott on just about everything.

While I was writing this book, it was known only as my "super-secret monograph." Most historians talk about the content of their projects and keep their methods to themselves. As an experiment, I decided to try doing the opposite. I refused to tell anyone what the book was about, but I shared my methods for researching and writing it. My thanks to all of the people who discussed the method with me in detail, and in some cases enthusiastically adopted parts of it: Torang Asadi, Chad Black, Dan Chudnov, Jim Clifford, Dan Cohen, Matt Connelly, John Fink, Marcel Fortin, Christopher D. Green, Mark Guertin, Steven High, Alan MacEachern, Eden Medina, Ian Milligan, Carolyn Podruchny, Matt Price, Matt Ratto, Elena Razlogova, Spencer Roberts, Geoffrey Rockwell, Jonathan Shaw, Bob Shoemaker, Stéfan Sinclair and Nathanael Smith. I hope they like the book, even though it doesn't come with a decoder ring after all.

I also owe a debt of gratitude to my students, colleagues, and teachers. While I was writing, I did reading courses on the history of the life sciences with Drew Davis and Michael Del Vecchio. Mikkel Harris tracked down and digitized obscure journal articles for me. Students in my big history course provided an abundance of enthusiasm and some thought-provoking questions. Colleagues at Concordia, York, and McMaster Universities, the University of Toronto, Code4Lib North, the Hertog Global Strategy Initiative at Columbia University, and the Digital Humanities Summer Institute in Victoria, BC, all discussed methods with me. Eric Fortune kindly e-mailed me about "the bizarre adventures that happen in lowland forests" while doing field work with electric fish.

Librarians are more important than ever in the age of digital resources: thanks to Liz Mantz and David Fiander at Western Libraries, and Marcel Fortin and Janina Mueller at the University of Toronto Libraries. Janina digitized all of the illustrations for the book. My editor at the Johns Hopkins University Press, Bob Brugger, and the reviewer for the press, J. R. McNeill, both provided excellent feedback for improving the manuscript, as did Glenn Perkins, the copy editor.

I first started learning about many of the topics in this book long before I decided to become a historian. At the University of British Columbia, Don M. Wilkie and Lawrence M. Ward taught me animal and human sensation, perception and cognition, respectively. At MIT, my SM supervisor, Steven Pinker, introduced me to a range of contemporary debates in evolutionary theory.

Big histories are necessarily based on the published work of others. Some historians make a point of fetishizing primary source research, but I have always felt like that signals a remarkable lack of trust in the abilities of our colleagues. If I can't rely on the work of a native speaker who immersed him or herself in a particular archival record for a lifetime, why should I think my own brief foray into unfamiliar territory would be epistemologically more secure? My debt to other authors should be clear from my citations, but a few deserve special mention: Mary A. B. Brazier, David Christian, Andy Clark, Stanley Finger, Leslie A. Geddes, Anita Guerrini, John L. Heilbron, Hebbel E. Hoff, Kevin N. Laland, Peter Moller, Iwan Rhys Morus, Laura Otis, Marco Piccolino, Simon Schaffer, and Kathlyn M. Stewart.

The Social Sciences and Humanities Research Council of Canada did not fund this book directly, but I am very grateful for their generous support for the hardware, software, and methodological research that made it possible. I am also grateful for support from the J. B. Smallman Publication Fund, and the Faculty of Social Science at the University of Western Ontario.

As with everything else, humanism has the potential to be practiced very differently in an electric world. This book was written entirely from digital sources, supported by a variety of computational techniques. These included automated searching, spidering, and concordancing, text mining (especially clustering), and machine learning. When sources were not already available in digital form, I digitized them myself or had them digitized before use. Many of the programs that I worked with are commercially available; when suitable software did not already exist, I wrote my own code in Mathematica. My notes on the method, which continues to evolve, are online at http://williamjturkel.net/how-to/.

And last, but certainly not least, my love and thanks to my wife and best friend, Juliet Armstrong. *Singe*!

# Spark from the Deep

# Introduction

We live in an electric world. Every moment of our lives unfolds within a vast assemblage of electrical systems, most of our own devising. These systems heat and cool our buildings, process, cook, and preserve our food; pump our water and waste; light the darkness; and power machines great and small. Electrical technologies allow us to see, hear, and act on scales that we can barely imagine, whether it is dragging individual atoms into place on a metal surface, listening to the background radiation of the Big Bang, or simply video-chatting with someone on the other side of the planet. Increasingly, and disconcertingly, we delegate capacities that once seemed ours alone—discrimination, communication, judgment, action—to systems that never tire, that can keep track of every detail and sense threats that we do not, that can be programmed to respond before we even know there is a problem. Despite gross global inequities, we now live in a world where electrical and electronic material culture are truly ubiquitous. These technologies have practically remade every aspect of the human experience within our lifetimes, and the changes pile up faster and faster. If we were somehow deprived of electrical systems for even a brief period, the global economy would collapse and the human population would crash.

We have not been denizens of this electric world for very long, as historians reckon time. The oldest widespread electrical technologies, such as artificial light, powerful batteries, electric motors, and wired communication, are barely a century and a half old. New electrical technologies have emerged at an ever-increasing pace since the nineteenth century: from radio and vacuum-tube electronics through the development of transistors and integrated circuits to the latest advances in computing. When cast in terms of proliferating gadgets, these developments appear to lie directly in the domain of applied physics. The discovery and peopling of an electric world is not solely of interest in the history of engineering or the physical sciences, however. The exponential growth of electrical technology has been paralleled by the steepest increase that the human population has ever undergone. And this demands explanation in terms that square with our understanding of human beings as a part of the natural world: as animals, in other words.

The knowledge and use of electromagnetism, electricity, and electronics is one of the defining features of our contemporary adaptation, and we have been taught to think of these things as artificial and abiotic. Could anything seem to have less to do with nature than an iPhone? This estrangement of electricity from life came about in the nineteenth century, however, and would have surprised many of the great thinkers of earlier eras. In most past times and places, the only accessible and replicable experience of powerful electric shock came from handling a few species of strongly electric catfish, rays, and eels. The oldest records that we have of human encounters with strongly electric fish portray in vivid terms the painful experience of being shocked by one. The names that people gave to these fish referred to their ability to induce torpor, trembling, or numbness. Everyone agreed that these unusual animals could discharge their electric shocks at will, stunning potential prey or would-be predators. Since the dead bodies of these fish were harmless, electricity was thought to be a property of life. The electrical discharges of the fish could be felt directly, or at a distance through water or through wet implements, but they could not be seen as visible sparks. Thus, there was no particular reason to link piscine electricity to inanimate phenomena like lightning or static electricity, although we now understand all of these occurrences to be the same kind of thing.

Beyond their ability to shock, people also found other ways to put the bodies of strongly electric fish to use, whole or in part. They ate them. They found therapeutic applications, such as using the shock of a live fish to treat a migraine or a prolapsed rectum. Later, electric fish became devices for experiment, demonstration, and entertainment. They posed some deep puzzles for Darwin because the theory of natural selection predicted that they had to have descended from ancestors who could only produce low-power electrical discharges. In the nineteenth century it was not at all clear what selective pressures would operate on an electric organ that that did not appear to be useful for feeding or defense. Subsequent generations of Darwinists have shown that, in fact, many fish have the ability to sense electric fields (as do a few other unusual animals like the platypus and echidna), and some fish can also generate weak electric signals to use for sensing and communication. The relatively recent discovery of these weakly electric animals and the ongoing effort to understand their full range of abilities has been spurred by continual advances in electronics, a field whose origins lie in the attempt to model animal electricity. In recent decades, electric fish have served as model organisms and sources of tissues, chemicals, and genes. Popular accounts of these animals tend to stress their potential to inspire biomimicry or biomimesis, as living examples that can be studied while engineering new sensors, computer interfaces, autonomous undersea robots, or energy-efficient batteries.

The central argument of this book is that our treatment of electric fish as *apparatus* enabled us to feel our way into electric worlds of our own and, eventually, to inhabit them. More generally, our evolutionary success is due in large part to the fact that we have the ability, perhaps unique, to treat our own bodies and those of other people and other animals as equipment. This, in turn, has helped us to expand on a scale that is far out-of-proportion when compared with any of our close kin. Less than a hundred thousand years ago, humankind numbered in the tens of thousands. With the development of agriculture the human population climbed to millions. We hit the one billion mark around the turn of the nineteenth century and had passed six billion by millennium's end. If we take the long view of biological time, the human discovery of an electric world has been abrupt indeed, but as with most sudden reversals of fortune, the origins of this dramatic expansion lie deep in the past.

The narrative challenge of studying electric animals as apparatus is to do justice to evolutionary, environmental, and historical processes, without artificially separating them. For this we turn to *big history*, a historiographical subfield that attempts to situate what we think we know about the human past in the wider context of what we think we know about the historical sciences such as cosmology, geology, and evolutionary biology. A key tenet of big history is the observation that similar *regimes* recur across different spatial and temporal scales. One such regime provides the theoretical framework for this book. In the literature of evolutionary biology, it goes by the name of *niche construction theory*; in environmental or ecological history, it is often simply referred to as the *dialectic*. Either way, the basic idea is the same. The activities of an organism modify the environment in which it lives, and this in turn changes the possibilities for its future.

Let's begin with evolutionary time. Niche construction refers to the fact that "organisms, through their metabolism, their activities, and their choices, define, partly create, and partly destroy their own niches." Since populations adapt to environments by a process of natural selection, any change in the environment can generate a form of evolutionary feedback. Until recently, however, those changes that are due to the organisms themselves were mostly ignored. There are plenty of examples. At the most basic level, many common animal activities, like hoarding food, migrating, creating fat deposits, sweating, or shivering, are adaptive because they have the effect of damping variations in resource availability. Animals also modify their environments. The activities of earthworms change the structure and chemistry of the soil in which they live, a process that Darwin himself studied. Many animals, including fish, select, modify, or construct microenvironments to shelter their offspring. Social insects create elaborate nests, and social rodents elaborate burrows. In both cases these shelters order the activities through which the group is maintained, regulated, and defended.

Evolutionary analysis that puts too much focus on the transmission of genes runs the risk of excluding these extra-genetic forms of inheritance. But if a gene (in conjunction with many other factors) disposes each individual that inherits it to modify its environment in a similar way, then environmental changes may be cumulative and thus "transmitted" to offspring. Note that such a mechanism does not necessarily imply a full-blown culture, which can be thought of as a more elaborate form of extra-genetic inheritance. The construction of niches also provides a mechanism for coevolution. This is because niche construction activities can affect the fitness, and thus the distribution, of genes in more than one population at a time.[1]

Ecologists are accustomed to thinking of interactions between organisms in terms of their trophic relations (who is eaten by whom). These food-web relations are governed by the conservation of energy and by principles of mass flow that readily lend themselves to a kind of accounting with well-known formulas. The construction of niches, however, results in the creation of less familiar *engineering webs* or *control webs*. An organism such as a tree may provide habitat or caching space for others. Trees exercise control over the abiotic resources that other organisms need, capturing water, sediment, and nutrients, affecting drainage, soil erosion, temperature, relative humidity, and the impact of rain and wind. These engineering or control factors are not nearly so easily accounted for as the more traditional measures of ecosystem interaction. Species that do the most significant engineering are not "keystone species" in a traditional sense, and their effects are still poorly understood. Nevertheless, the potential significance of this way of thinking for ecological history and for the history of science and technology should be clear. All humans, including the so-called people without history, have agency and technologies and modify their environments constantly.[2]

When we take the much shorter time frame of deep history, and focus on the hominin lineage, we see humans engaged in an endless process of experimenting with other organisms. This is another notable example of niche construction. The details of when and where particular plants or animals were first domesticated have been studied for decades. But these domesticates did not arise in a vacuum. They were the result of human activities like selective burning, weeding, watering, transplanting, confinement, culling, breeding, and the like, undertaken in response to larger-scale pressures such as population increase or climate change. These activities were also integrated into a behavioral context of management and manipulation of a broad range of species and habitats. Domestication and agriculture originated independently as many as ten different times in different parts of the world, and each took a different trajectory. This path dependency is exactly what we would expect from a dialectical phenomenon. Later, the domestication of plants and animals gradually led from nomadic to more sedentary lifeways, and this, in turn, resulted in the construction of permanent dwellings, the establishment of agricultural fields, boundaries, watercourses for irrigation, fortifications, roads, and so on. Each of these could far outlast their creators, establishing "more favorable conditions for selection on niche construction" and raising "the intriguing question of the extent to which humans have been shaped by natural selection to behave in accordance to their impact on future generations, be it at a local or at a more global scale."[3]

In this volume, I explore some significant consequences of a uniquely human behavior over the lifetime of our kind. Although niche construction occurs across a wide variety of species, the habit of experimenting with other organisms is distinctly human. We disassemble, modify, remix, and reassemble the bodies of ourselves and other organisms to instrumental ends. Our story touches on practices of butchery, dissection, autopsy, medical experimentation, and technological enhancement. Throughout the human career, people have pursued and sustained electric encounters with an increasing number of organisms—not only fish, but eventually frogs and dogs and others—because doing so has allowed them to accomplish goals that they could not satisfy in any other obvious way. These episodes were usually puzzling, often painful, sometimes cruel, and occasionally downright gruesome. They were, in all the senses that we use the word, shocking. In order to trace the currents of this history, we have to maintain contact with the sites where these electric exchanges occurred, typically across the seams of human, animal, and machine. Our story is thus fleshed out from slabs of the past that are often kept apart: medicine and engineering, electromagnetism and evolution, animals and machines . . . stitched together and animated with a spark.

ONE

# Strongly Electric Fish

## The Earliest Encounters

The strongly electric African catfish *Malapterurus electricus* inhabits the River Nile and most of the freshwater lakes and rivers south of the Sahara, from the Senegal and Niger Rivers to the Zambezi. It feeds on other fish, usually just after sunset, stunning its prey with a volley of electric shocks that can last for several seconds. It has taste buds embedded in the skin of its entire body, as well as in its mouth, and it can detect prey with any of these. *M. electricus* swims slowly and feeds indiscriminately. It is content to paralyze and eat large numbers of small schooling fish, like herrings and cichlids, and small numbers of larger fishes, worms, insect larvae, and crustaceans. Occasionally it even eats mud and plant matter. It grows quickly to over one meter in length, taking on the form of a "rather rigid sausage" in grey or brown, with a pink tint and freckles. Like other catfish, its mouth is fringed with barbels that resemble whiskers; unlike others, it does not have spines or armor, presumably because its ability to unleash 300–400 volts at will is defense enough. Nocturnal, it prefers to lurk among roots or rocks in standing or slowly moving water near the shore, and breeds in underwater caves or shelters. In the Congo and Niger basins, *M. electricus* shares its habitat with two other species of electric catfish that are quite similar. In the nineteenth century, West African electric catfishes were assigned to a variety of different species and genera, but these have gradually been found to be synonymous with *M. electricus*. Researchers suspect, however, that there are other species of electric catfish in West Africa that have yet to be recognized.[1]

The first pictorial and written depictions of strongly electric fish come from Ancient Egypt, but our relationship with these animals is much older. Freshwater catfishes have been in Africa for at least 100 million years, but they are also found throughout South America, eastern North America, central Europe, and South, East, and Southeast Asia. Since the ability of *Malapterurus* to discharge strong electric shocks is not shared with catfish elsewhere, it must have evolved after continental drift separated Africa from South America. For unlike birds (which can fly), plant seeds and spores (which can drift in the wind or float on the water), and land animals (which can swim or be transported by flotsam), freshwater fishes are constrained to remain in bodies of fresh water to survive. The only way they can travel any distance is when those bodies of water flow into one another. Exactly when some African catfish became strongly electric is not currently known, but the best estimates suggest that it was between 25 and 85 million years ago, long before the hominin lineage emerged. Electric catfish inhabited all major freshwater systems in Central and East Africa, and catfish fossils have been found in regions where the fossils of hominins have also been discovered. So long before they were properly human, our ancestors must have been interacting with strongly electric catfish, even if their encounters were mostly limited to scavenging the dead ones and avoiding being shocked by the live ones. After all, human beings need to obtain drinking water more or less every day, and thus we have had to live near freshwater habitats for most of our existence as a species.[2]

Although we have no written records describing early encounters between people and strongly electric fish, we need not be discouraged. We have no written documentation for the origins of pet-keeping, animal domestication, swidden agriculture, the first zoonotic diseases, or practically any other longstanding relationship between humans and other organisms. Writing, after all, was invented late in the human story. Instead,

The strongly electric catfish *Malapterurus electricus* was common in the aquatic environments that supported early hominins, and it has remained well known to African cultures. From Georg Ebers, *Egypt: Descriptive, Historical, and Picturesque*, vol. 1, translated by Clara Bell (London: Cassell, 1887).

we can make use of evidence and inference from a range of historical sciences, following the tenets of big history. These scientific studies highlight the importance of aquatic environments for early hominins, demonstrate an extensive process of niche construction, and reveal the origins of the human capacity for regarding people, animals, and inanimate objects alike as equipment. An understanding of each of these factors is crucial to seeing not only how encounters with electric fish eventually inspired people to inhabit an electric world but also how the details fit into a wider account of what it means to be human and what it means to be alive.

The last animal who was an ancestor to both apes and humans lived about seven million years ago; estimates made from molecular systematics suggest that it lived more recently than that, while those made from paleontology put the event further back in time. After that point, whenever it was exactly, the hominin lineage could be said to have emerged. The fossils of our ancestors reflect a process of adaptive radiation: the exploration, by natural selection, of a range of possible ways to be hominin, rather than a "single-minded slog from primitiveness to perfection." This adaptive radiation was likely the consequence of environmental change. Prior to about 10.5 million years ago, Africa was heavily forested, and the more distant primate ancestors of apes and humans were adapted to this arboreal habitat. They had front-facing eyes with stereoscopic vision and good depth perception, relatively large brains for processing complex three-dimensional information, and dexterous hands and feet for gripping and climbing. Cooling global climate and a decline in seasonal equatorial rainfall allowed regions of savanna to break the African canopy, creating a new patchwork of habitats at the edges of open woodlands and grasslands. Several African mammalian lineages show extinction and speciation in the wake of this environmental variability. Some of the hominids, perhaps larger, with larger brains and better manual dexterity, were in a position to exploit new possibilities. Selective pressure in these novel habitats favored anatomical adaptations that allowed more habitual bipedalism.[3]

The hands and arms of the first bipedal hominins retained features that allowed them to climb and to continue to forage in trees, although not as efficiently as apes, and their legs allowed them to walk upright, although not as efficiently as we do. Their brains were small, less than a third the average size of our own. Nevertheless, they were quite successful, persisting for several million years in the expanding edge habitats of African forests. Their teeth were adapted for grinding plant matter, but they would have eaten meat when they could get it: insects, lizards, birds, or carrion. This is true of most living primates. They would also have regularly encountered freshwater fish while drinking or foraging at the edges of rivers. There they also discovered that they could make stone flakes by striking a river-rounded cobble with another stone at just the right angle. These sharp-edged flakes were the earliest stone tools, and were used for a variety of cutting, butchering, and skinning tasks. Although tool use would become one of the outstanding hallmarks of our own genus, *Homo*, the first stone-tool makers had to have been individuals who were not all that different from their own parents. In other words, the toolmaking breakthrough must have occurred within one or more of the australopithecine species rather than signaling the sudden emergence of an entirely new species.[4]

Early hominins were characterized by an array of anatomical adaptations that increasingly distinguished them from apes. Their brains and bodies were getting bigger while their teeth and guts got smaller. These changes are assumed to have followed from an increasing dietary dependence on nutrient-rich food sources, like animal flesh, nuts, or buried roots. The implication is that early hominins were thus becoming more dependent on tool manufacture and use. Early stone tools have often been found in conjunction with animal bones, including those of mammals, crocodiles, turtles, and fish. Researchers have tended to concentrate on the mammal bones, consciously or not, perhaps because the act of killing large mammals best fits the impression of "man the hunter." But there is good reason to suspect that fish were an essential component of the hominin diet, too.[5]

One line of evidence comes from studies of present-day hunter-gatherer communities in Africa, like the Hadza and San, who kill an average of one or two medium-to-large mammals per week. While this is significant, it is not enough to feed fifty or so people, and during periods of drought or nutritional stress, the situation worsens. Meat from game animals becomes harder to obtain and too lean to rely on as a primary food source, so alternatives must be sought. Plant foods, especially nutrient-rich underground tubers and roots, are now thought to have been key "fallback foods" for early hominins. These underground storage organs are most dense in shallow-water or flooded habitats and remain nutritious during drought. Most hominin fossil sites are associated with lake margins, floodplains, or streams, where these foods can reliably be found. Wild baboons have been observed to make intensive use of these plants during the dry season. Not coincidentally, the murky and still water of lacustrine reedbeds also serves as habitat for fish, including catfish. Baboons also eat live fish from the water and scavenge newly dead ones from the shore, both during the dry season, when fish are trapped in evaporating pools, and during the rainy season, when they are spawning. What works for baboons presumably worked for ancestral hominins, too.[6]

Extensive fish remains have been found at early hominin fossil sites. The vast majority of bones are from the (non-electric) catfish *Clarias*, which can grow up to 2 meters in length. *Clarias* also dominates later hominin assemblages, along with *Barbus*, a large minnow-like fish. *Barbus* usually cannot be caught bare-handed, but they are easily clubbed or speared in pools while spawning. Some of these fossil sites long predate modern humans. Cut-marked catfish bones were recently found at a site in northern Kenya that has been dated to 1.95 million years ago, along with the similarly marked bones of turtles, crocodiles, and hippopotamuses. These bones provide additional proof that hominins made use of a more diverse diet than just big game and that they opportunistically exploited aquatic resources as well as terrestrial ones.

Catfish in rivers are difficult to catch without equipment, but they spawn in floodplain waters that are only a few centimeters deep. Contemporary ethnographic accounts show that traditional fishers in Africa can catch catfish with their bare hands in large numbers during spawning. Returning downriver after spawning, the catfish become stranded in shallow pools by the hundreds as the dry season advances, giving fishers even more opportunity to catch them. Since this is also the time when plants and terrestrial mammals are under the most stress, it would have been advantageous for hominin groups to focus on fishing in the late dry and early wet season and to organize their subsistence rounds to coincide with the best opportunities for harvesting the fish. Among present-day hunter-gatherer communities, such occasions often temporarily bring together larger groups, resulting in new opportunities for communication and social exchange. The fossils thus tell a story of millennia of seasonal exploitation of fish by hominins, especially catfish. In taking these animals, our ancestors would have occasionally been shocked by *Malapterurus*, something that would have made a strong impression on them.[7]

Fish protein is comparable to that obtainable from meat, and eating fish also provides essential vitamins, minerals, fat, and oils. Catfish are particularly sought in present-day Africa as a good source of fat and oil. Current debate focuses on the degree to which aquatic resources were necessary for the expansion of brains in the hominin lineage, as fish and shellfish are a much better source than meat for some nutrients that may be essential for brain development. Whether or not our bodies are capable of synthesizing these nutrients (docosahexaenoic acid and long-chain polyunsaturated fatty acids) is an active area of research.[8]

## The Mind behind the Tool

Later in the book we will explore the question of how fish evolved the ability to discharge electric shocks and to sense electric fields. Human beings now have converged on similar electrical abilities and have discovered many new ones. Although not often framed in evolutionary terms, our electrical abilities owe as much to our biology as those of fish do to theirs. The mechanisms in our species are very different, of course, but that does not make them any less natural. We can begin to find the origins of these mechanisms in hominin fossil and tool assemblages. One line of evidence comes from taphonomy—the study of the decay and possible fossilization of organic remains—and, in particular, from the study of butchery. Our ancestors originally learned how to skin and disarticulate other animals efficiently in order to feed on them, and the skills and tools of butchery informed later practices of dissection. Animals could be disassembled to provide food for the mind as well as the body. The construction and use of the tools themselves also have much to teach us about hominin cognition, both individual and social. The steps required to fashion and use a tool, the gathering of necessary materials, and the repurposing of existing activities toward new ends all reflect a particular kind of mind: one that imagines itself surrounded by *apparatus*. Recent research confirms the fact that many different kinds of animals make and use tools; people are not the only tool-using animals by any means. But no other animal entirely surrounds itself with a varied material culture of its own devising. This is the human niche.[9]

First, hominin use of other animals. Early hominins were traditionally portrayed as big-game hunters, but in recent decades paleoanthropologists have focused on other strategies like scavenging. As Ian Tattersall notes, hominins may have been able to steal from the carcasses that leopards cached in trees. Or they might have used their tools to cut small pieces from the carcasses of already dead animals, carrying the food someplace safe to eat. Hominin individuals scavenging or hunting on the African savanna were certainly not alone there. They faced competition and predation from leopards, lions, hyenas, raptors, and other animals. Even if the flesh from a dead animal was already gone, hominins cracked the long bones with cobbles to get at the marrow.[10]

Fossil remains can provide some evidence for hominin practices of scavenging or butchery, but they need to be interpreted with caution. In the absence of human intervention, East African mammal remains undergo a predictable sequence of decay. After the animal is killed and torn apart by lions or hyenas, the carcass is then scavenged by jackals or vultures, eaten by insects and bacteria, weathered by heat and moisture, and trampled by other organisms. Despite the number of different processes that might be involved, the order in which mammalian skeletons come apart is generally consistent across a broad range of species: the looser joints separate first. Forelimbs and tail come off the body, the jaw separates from the skull, and so on. The last joints to disarticulate are the vertebrae of the neck. These findings give us a baseline to compare with anthropogenic butchering. In a contemporary meatpacking plant, an animal carcass is sliced into consumable parts using electric power saws. These cut through bone easily, resulting in chops and steaks in the supermarket that still have bone segments in them. In the absence of power tools, it is much easier to cut around the bones and through joints, breaking the bones only to extract marrow. People without power tools thus follow the natural order, separating loose joints first. They make enough easy cuts to convert a carcass to manageable proportions, then stop cutting. This means that the bone assemblages found with stone tools can be interpreted as evidence for hominin butchery, but only with caution, since similar patterns of bone dispersal occur in the absence of people.[11]

Hominins extracted many useful animal products besides meat and marrow, including fat, blood, brains, entrails, hide, sinews, and tendons. Each of these needed to be extracted from the carcass in a particular order, resulting in characteristic cut marks on bone. Most researchers would agree that the fact that there are so many animal bones from different taxa that have cut marks on them—"literally from hedgehogs to elephants"—is an indication that toolmaking hominins were getting more meat from animal carcasses than do present-day chimpanzees or baboons. Exactly what balance of meat eating was the result of scavenging, hunting, or some other strategy remains to be determined. A site where stone tools were found with the bones of a single animal might be interpreted as a butchery; a site where tools were found with the bones of many different animals might be interpreted as a camp to which dead animals had been brought for butchering. These latter sites might in turn indicate some division of labor into hunting and gathering, and might reflect practices like food sharing.[12]

Second, the tools themselves. Archaeologists replicate stone tools to gain insight into their manufacture and use. Experiments with re-created stone tools show that flakes with unworked edges (such as the earliest stone tools) are quite good for cutting muscle. They also show that different species of animals require different relative amounts of effort to cut muscle and sever joints. Bifaces, sharp-edged tools that are created by symmetrically removing flakes from both sides of a stone blank, tend to be more suited to tasks like skinning. So it is clear that a stone tool kit can be specialized for the kinds of animals that are likely to be encountered, and the kinds of extractive tasks that the tools will be put to. For several thousand years, members of *Homo* created and used the same kinds of unworked stone flake tools as their australopithecine forebears did. Starting about 1.5 million years ago, however, they began to create a wider variety of tools, including hand axes, picks, and cleavers. These were painstakingly worked, requiring a substantial investment of effort and placing new kinds of cognitive demands on the creators.[13]

This point is best appreciated by comparing these new kinds of stone tools with the products created by other animals. Many species build elaborate structures to mediate their interactions with the environment and with other organisms: spider webs, beaver dams, and bird nests come to mind. These structures have a relatively simple algorithmic description: *add something, repeat*. The unworked flakes of australopiths can be described in similar terms: *remove something, repeat*. Although it can be difficult for novice knappers to get the angle of blow just right to create a sharp-edged flake, we do not need to assume that australopithecine toolmakers were trying to impose any particular form on their products. (They may have been, of course.) The new stone tools of *Homo* were a different matter, as they show a significant degree of imposed form and are sometimes nearly identical to one another. To create a biface, hominin knappers had to imagine the final form, keeping it in mind while they were working, and adjusting it as necessary to deal with unforeseen contingencies such as flawed material. An algorithmic description of such a process is considerably more complicated than that required for the earlier unworked tools, as is a mind capable of executing it.[14]

When we compare hominin transport practices with those of chimpanzees, we find further evidence that toolmaking hominins were beginning to think in more complicated ways than their ancestors or kin. Wild chimpanzees in an Ivory Coast park use hammerstones to crack two species of nut, one with a harder shell than the other. Primatologists discovered that the chimpanzees would transport heavier hammerstones for longer distances to crack the nuts with the harder shells. The chimpanzees also remembered the location of various hammerstones and would choose to retrieve the stone closest to the tree where they were feeding—sometimes from a distance of several hundred meters away. In order to do this, the chimpanzees must have a map-like mental representation of their territory, in which they can locate themselves and compare the relative distances between stones, trees, and their own current position.

They are also apparently able to make decisions about the relative costs and benefits involved in choosing a particular tool for the task at hand. In other words, chimpanzees are able, at least to some extent, to *imagine* an outcome rather than simply trying an activity to see how it turns out. The researchers also showed that young chimpanzees learn how to crack nuts by observing adults, and that wild chimpanzee behavior leaves an archaeological record of transported stones, stone fragments and plant refuse, an assemblage that can be compared to those of early hominins.[15]

Early hominins displayed even more sophisticated transport practices than those of contemporary chimpanzees. These reflect an imaginative capacity closer to, although perhaps not identical to, our own. For one thing, some hominin sites have animal bones with cut marks on them made by stone tools, but the tools themselves were not found at the site. This implies that the hominins brought the tools to the butchering site, then carried the tools away with them when they were finished, along with the meat. In addition, at many sites with hammerstones, cores, and flakes, there are stones that could only be present in the sediments if they were carried there by hominins. This suggests a degree of foresight: the hominins anticipated that they would need tools and brought the raw materials for toolmaking with them. By regularly carrying around tools and raw materials, hominins distanced themselves from the tool behavior of other animals, moving closer to our own idea of possessing *equipment*. Other animals use artifacts infrequently, if at all; contemporary human societies have an all-encompassing *material culture*, which mediates nearly all of our interactions with our environment and with one another, from birth to death. We started along this path about 2.5 million years ago. The debitage created during episodes of knapping had the effect of "radically alter[ing] the environment of opportunity . . . The abandoned hominin hammers, cores, and flakes became a potential, and ongoing, new resource for the same or different individuals, in effect constructing a new niche for them."[16]

Human cultures often recruit existing activities to serve new functions, but this is not true of other animals. When wild chimpanzees use stones as hammers, they occasionally break off sharp flakes, but there is no evidence that they use the flakes themselves as tools. If the accidental flaking of hammerstones during pounding led hominins to create flakes intentionally by percussion, then we have evidence for the beginnings of one specifically human adaptation. This process took several hundred thousand years at least, and some tools, like hand axes, remained essentially unchanged for a very long time. People would later find stone knives and spear and arrow points to be very useful, which implies that our earlier ancestors would also have profited from having them, if only someone could have imagined their creation. The absence, too, of carved stone or bone objects like statues, pendants, and beads suggests that early hominins were still unable to adapt activities fluidly from serving one function to another. But they had begun to modify their environment in such a way that it provided new opportunities for action—what the psychologist J. J. Gibson called "affordances"—and ongoing exploration of these affordances would eventually lead to an explosive increase in artifact production and use.[17]

A complementary way to think about the significance of stone technology is to see that hominins were developing an ability to treat parts of the world as an extension of their minds. For a hand ax is not merely a tool, but also a template for how to manufacture more hand axes, and it is a template that is available to everyone in the group. Artifacts could become a part of the hominin mind, in other words, without being part of any particular individual's brain. Iain Davidson and William C. McGrew write, "In all of the examples of cultural behavior among wild non-human primates, there is no other example where new behavioral opportunities opened up as a consequence of prior ones. Discarded nut shells are just that, and not a resource for some other activity." Knapping and butchery, however, created a niche with many new possibilities.[18]

The construction of niches creates both a supply of potential technologies and a demand for them. On the one hand, raw materials, artifacts, byproducts, activities, and processes can all be repurposed or recombined in new forms. On the other, opportunities for the adoption and use of technologies are provided by the ramification of human and technical needs. Once we accept that technologies must be created from preexisting ones, then it follows, as the economist W. Brian Arthur writes, that "every technology stands upon a pyramid of others that made it possible in a succession that goes back to the earliest phenomena that humans captured." Rather than simply attempting to determine ultimate causes, this concept helps us to understand the contingency, path dependency, and dynamics of the evolution of technologies. For there is a dialectic at work here just as in environmental or ecological history: the choices that we make shape our future opportunities and help to determine what we become. From the perspective of big history, we will see the same regime operating on temporal orders ranging continuously from the very long term to the very brief. This is the perspective that we need to take if we are to understand how we were inspired by other animals to colonize an electric world and what that, in turn, tells us about human nature.[19]

## Exploiting Aquatic Resources

Hominins are able to walk without stopping for long periods of time, and this ability took them from East Africa into the rest of the continent, and beyond. The first wave of migrations began about 2 million years ago, and carried hominins as far as Java within a few hundred thousand years. People entered environments where they could exploit new resources. Since we can assume that they traveled close to fresh water whenever possible, many of these must have been aquatic. Until the recent discovery of the cut-marked catfish bones in Kenya, most of the evidence for early hominin consumption of aquatic food sources was inferential. Java, for example, was at the time full of extensive rivers, lakes, swamps, and marshes with more than a dozen edible species of fish and mollusk that could be harvested without elaborate technology. These included the Asian redtail catfish *Hemibagrus nemurus*, which spawns in shallow water and can be clubbed there, and two species of "walking" catfishes, *Clarias batrachus* and *C. leicanthus*, which can breathe air through auxiliary organs and thus swim in very shallow water and even "crawl" overland for short distances. These fish can easily be caught by hand during the dry season and are enthusiastically eaten by present-day human groups. In fact, seasonal consumption of aquatic organisms appears to be common in nonhuman terrestrial mammals and is documented for a wide variety of predators ranging from mice and rats to wolves, bears and baboons. If our hominin ancestors did not behave similarly, we would have to explain why not.[20]

Our own species, *Homo sapiens*, emerged in Africa between 150,000 and 200,000 years ago. After that time, the regular use of aquatic resources becomes much more apparent in the archaeological record. In

Ethiopia, modern human skeletal remains dating from about 100,000 to 200,000 years ago were found with stone tools and large numbers of mammal, bird, and fish bones. Most of the birds were water birds like herons, pelicans, and darters; the fish were predominately catfish and Nile perch (*Lates niloticus*). The presence of barbed bone points at the site suggests a long history of spearfishing in the region. Similar barbed points appear at other sites in eastern and southern Africa, often associated with catfish remains. These barbed points resemble the spines of many species of catfish, leading paleobiologist Josh Trapani to suggest that human fishers may have been inspired to create the artifacts after painful experiences with non-electric catfish.[21]

Hominins also learned to exploit the resources of the seashore. The oldest evidence for human adaptation to coastal environments comes from cave sites in South Africa and around the Mediterranean. In South Africa, marine shellfish middens have been found that date back as far as 164,000 years ago. These also contain ostrich eggshells, tortoise remains, and stone tools and came over time to include new species such as limpets and mussels. These people did not fish in the sea regularly, however, as neither representative fish bones nor net weights have been found. Unlike freshwater catfish, the marine fish in coastal environments were difficult to harvest without elaborate equipment. Shellfish, in contrast, could be harvested by almost every member of a hominin society. Coastal environments also served as places of refuge and corridors for movement during periods of glaciation. During one such episode, between 123,000 and 195,000 years ago, the breeding population of *Homo sapiens* dropped precipitously, leaving hundreds, or at most thousands, of humans huddled around the edges of the ocean. Fossil remains from South African coastal sites show that human populations there survived on shellfish and a highly diverse collection of underground storage organs, both of which were relatively abundant food sources during cold and arid conditions.[22]

Humans responded to a hostile world by developing ever more elaborate technologies. The divergence of modern humans was followed by a wave of global migrations beginning around 70,000 years ago and continuing to the present. In the tens of thousands of years before the modern human expansion, new behaviors were gradually being explored and adopted piecemeal at a variety of sites in Africa. In addition to intensive use of aquatic resources, these included increased range and more extensive long-distance trade; specialized hunting techniques; heat-treated stone tools; an increasing diversity of artifacts, including hafted, composite, and bone tools; and the mining, manufacture, and use of pigments. From the time of African emigration there is unequivocal evidence for the kind of innovation that we associate with modern humans. At that point there are objects that were clearly intended to be art; artifacts made from bone, shell, ivory, and antler; remnants of inhabitable structures; bone flutes; graves that suggest ritual burial; and many other kinds of evidence for cultural innovation. Fossil bone assemblages also show that fish other than catfish were being routinely consumed, suggesting that fishing technologies continued to improve.[23]

The evidence for early human fishing remains tantalizingly difficult to assemble. The technologies that allowed intensive fishing, such as nets, harpoons, boats, and weirs, depended on more general techniques like weaving and woodworking. The timeline for the emergence and development of these activities is constantly being revised, for several reasons. Many of the artifacts would have been made from materials that cannot survive for long periods in the fossil record (unlike stone). Fish bones tend to be less dense and more fragile than those of terrestrial mammals and are thus less likely to be preserved. Sharks, rays, sturgeon, and some other edible fish have a cartilaginous skeleton, which does not fossilize well. The predatory activities of nonhuman animals create bone assemblages that can be confused with anthropogenic ones. The rise and fall of sea and lake levels associated with periods of climate change has a deleterious effect on coastal sites, and much of the relevant evidence is thought to be currently under water. In some areas like northern Australia, the coastline has moved laterally by as much as 1,000 kilometers over the past 20,000 years.[24]

Until the last decade or so, there has been something of a paradox in the anthropological and archaeological research literature. Anthropologists have long recognized that some of the most sophisticated, populous and affluent hunter-gatherer groups live on coasts and are supported primarily by aquatic resources. The aboriginal cultures of the north Pacific are a prime example, but there are others, like the Gunditjmara people of Australia, who practiced an elaborate eel aquaculture. At the same time, there has been reluctance in some quarters of archaeology to acknowledge the importance of adaptation to aquatic environments in human evolution. In the late 1960s, C. W. Meighan speculated that "humans apparently existed for uncounted millennia before that anonymous hero ate the first oyster," and S. L. Washburn and C. S. Lancaster described water as "a barrier and a danger, not a resource" for early hominins. But such a hydrophobic attitude is hard to square with the evidence that modern humans made several long sea crossings while colonizing insular Southeast Asia and greater Australia. This happened long before the last ten or fifteen thousand years. It is possible that *Homo erectus* made short sea crossings in Southeast Asia as long as 800,000 years ago.[25]

In retrospect, it is clear that aquatic faunal remains have been a consistent feature of hominin sites around the world dating from at least 2 million years ago. Some of the sites outside of Africa long predate modern humans. At Kao Pah Nam, in Thailand, freshwater oyster shells were found in a cave alongside a hearth, stone tools, and bones from terrestrial animals, dated to 700,000 years ago. Hippopotamus and freshwater turtle remains were found in Israel in an assemblage consisting mostly of land mammals, dated to 400,000 to 500,000 years ago. An English site from about 300,000 to 350,000 years ago contained the remains of freshwater fish, beaver, otter, and waterfowl. Multiple sites in France, dated between about 150,000 and 300,000 years ago, contain marine shellfish. Sites in Morocco, Algeria, Libya, Gibraltar, and Italy contain the remains of limpets, mussels, crabs, clams, and fish associated with hominin settlements dating between about 50,000 and 130,000 years ago. Aquatic adaptations are even better documented during the global expansion of modern humans. The oldest evidence of complex fishing technology are barbed bone harpoon points found at two sites in Zaire, along with the remains of thousands of large freshwater fishes, dated to about 80,000 years ago. Later sites elsewhere in Zaire, Upper Egypt, and Egyptian Nubia also show sizable numbers of freshwater fish bones. We can be certain that ancestral African hominins had extensive experience with the strongly electric catfishes of the genus *Malapterurus*: after all, their habitats overlapped for more than two million years. As hominins made ever more use of coastal marine resources, they would have sooner or later encountered another strongly electric fish, the torpedo, or electric ray.[26]

Torpedoes inhabit all of the major oceans. Like other rays, they are flattened, disk-shaped fish with enlarged pectoral fins. They have thick, finned tails and fleshy bodies. In the Mediterranean there are three different species of electric ray, all of which are capable of delivering electric shocks of up to 200 volts or more. The smallest of these is *Torpedo torpedo*, an orangish-brown fish about 60 centimeters in length, with five blue spots on its upper surface. It can typically be found at depths ranging from 2 to 70 meters, preferring tropical inshore waters with soft bottoms. Its range extends to the Atlantic coasts of Europe and Africa, from the Bay of Biscay in the north to Angola in the south. *Torpedo marmorata* grows to about a meter in length and is dark brown and marbled in appearance. It is also found inshore, although sometimes at greater depths than *T. torpedo*; it ranges farther, too, being found along Atlantic coasts from the United Kingdom to South Africa. It also inhabits the eastern coast of Africa and the coasts of the Red and Arabian Seas as far as India. *Torpedo nobiliana* is the largest of the three, growing to a length of 1.8 meters, and has the most extensive range. It is found along coasts in the western Atlantic from Nova Scotia to Brazil and in the eastern Atlantic from Scotland to South Africa. *T. nobiliana* is uniformly dark in color, shading from chocolate to shiny black. Although torpedoes can be found in estuaries and off river mouths, they will not enter fresh water. They are nocturnal, using their electric organs to stun the benthic fish and crustaceans that they feed on. By day they bury themselves in the sand so that only their eyes and spiracles—gill slits behind each eye—are exposed. Lurking beneath the sand, torpedoes are occasionally stumbled over by people wading near the shore, an experience that reportedly feels like "being hit by a very large fist."[27]

An assortment of other torpedoes have been identified in the oceans of the northern hemisphere. These may belong to separate species, or they may be regional variants of the ones listed above. In the southern hemisphere, torpedoes are rare, and the related numbfishes predominate instead. These may be found from the east coast of India to the coasts of Japan and Australia in the Pacific. As with torpedoes, all of these fish are able to generate strong electric shocks, to the dismay of people who step

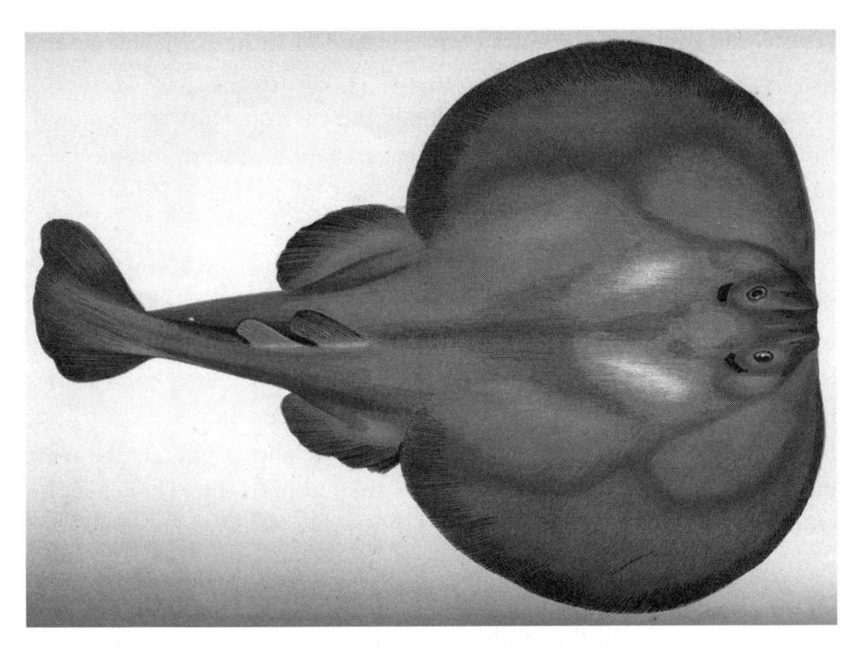

The strongly electric marine *Torpedo* was the subject of experiment in the Mediterranean region from classical antiquity onward. From Jonathan Couch, *A History of the Fishes of the British Islands* (London: Groombridge, [1867]).

on them accidentally. Note that stingrays, although related to electric rays, are not electric. They sting with a venomous barb located on the tail rather than with electricity.[28]

The modern humans who emigrated from Africa left behind the freshwater habitats of the strongly electric catfishes. But those people who traveled or remained near temperate or tropical oceanic coasts—now thought to be a significant proportion—had increasing opportunity to encounter strongly electric rays. This would have been a consequence of harvesting intertidal resources like shellfish, of the increased use of nets and other fishing technologies, and of the use of rafts or boats. Fossil evidence for human-ray encounters is scarce: as cartilaginous fish, rays are even less likely to leave traces in the archaeological record than bony fishes like catfish. Nevertheless, there is evidence for aquatic adaptations at a number of sites in the Pacific, including many shellfish middens, and the human colonization of those places required extensive sea voyaging. Boats must have been in use in insular Southeast Asia, Australia, and

New Guinea by about 50,000 years ago and in western Melanesia and the Ryuku Islands (between Taiwan and Japan) from 35,000 to 40,000 years ago. In every case, humans altered the environments they entered. But marine animals tended to be more resistant to anthropogenic pressures than land animals because they could take refuge offshore or in deep waters.[29]

The colonization of the New World from Siberia and Beringia began during an Ice Age. The ancestors of American aboriginal peoples had lived in a cold climate for millennia, and probably skirted glaciated regions by traveling along resource-rich coasts. They would have explored routes into the interior of the continent whenever these were not blocked by ice, but interior regions would not have been able to support nearly the same number of people as comfortably as the Pacific coasts did. These peoples would probably not have had cultural memory or personal experience with strongly electric fishes. They would encounter some new species in relatively short order, however. On the Pacific Coast between present-day British Columbia and Baja California, relatively large, dense, and sedentary hunter-gatherer societies shared their habitat with the marine electric ray *Torpedo californica*. Occasionally aggressive toward human divers, this torpedo delivers a shock "strong enough to knock down an adult."[30]

One place where these encounters were almost sure to have occurred is the Channel Islands of California, which were colonized about 13,000 years ago. Extensive kelp forests grew near the island shores, providing food and habitat for abalones, mussels, snails, sea urchins, and other shellfish. These, in turn, were preyed upon by lobsters, birds, sea otters, and people, who were, naturally, happy to eat the other shellfish predators, too. Since the Pacific electric ray prefers to inhabit kelp beds and sandy bottoms and is found in depths as shallow as 10 feet, we can assume that it was well known to coastal aboriginal peoples. Abalone shells and sea otter bones appear in island archaeological sites in good numbers, along with artifacts and fish and bird remains. A study of 11,000 abalone and mussel shells from these sites shows that their average size declined progressively over the last 10,000 years, evidence for the increasing impact of human predation. South American sites in Chile, Peru, Ecuador, and Brazil dated between 12,000 and 9,000 years ago also show extensive evidence of shellfish, marine and freshwater fish, and seabird exploitation.[31]

When human colonists moved into the freshwater river basins of the northern part of South America, they discovered an entirely novel and frightening animal: the electric eel. Despite its common name, *Electrophorus electricus* is not an eel but rather the largest of the gymnotiform fishes. It grows to a length of up to 2.5 meters and a weight of more than 20 kilograms. It has a blunt head and is roughly cylindrical in shape, in contrast to other gymnotiforms, which are much narrower laterally—hence their common designation as "knifefishes." The electric eel is dull grey, with yellow, red, or black coloration on the ventral surface of its head and belly. Its back is finless; it swims by undulating a single long fin that runs the length of most of its underside. Christopher W. Coates, the director of the New York Aquarium in the mid-twentieth century, noted that "two absurd little round fins stick out on either side of its head like the ears on a cartoon character." But the electric eel is not to be trifled with. The longer the fish, the higher the voltage it is capable of discharging. A 1.5-centimeter larva generates a few dozen millivolts when feeding or disturbed. A juvenile, 7 centimeters in length, is already capable of generating almost 100 volts, and the discharge of a fully grown adult can be over 600 volts. *E. electricus* inhabits calm, muddy waters and is frequently found in swamps, creeks, flooded forests and coastal plains, where it can be one of the predominant aquatic predators. The electric eel is nocturnal, but it can be seen swimming in open water in the late afternoon, in search of small fish to eat.[32]

Humans evolved in an environment where strongly electric catfish were regularly encountered, and the global expansion of humankind brought members of our species into close contact with two new kinds of strongly electric fish: marine rays and freshwater "eels." By the time that people first encountered strongly electric rays and the electric eel, humans had long occupied niches that were largely of their own making. They had an inherent habit of experimenting with a broad assortment of other species of plant and animal wherever they went, and they would have been familiar with strongly electric fish when their habitats overlapped. As with every aspect of human environments, the fishes' ability to shock was something that might be put to other uses, a resource for niche construction.

Despite its common name, the freshwater *Electrophorus electricus* of South America is not an eel. It is the most powerful of the strongly electric fish. From Hermann J. Meyer, *Meyers Konversations-Lexicon: Eine Encyklopädie des allgemeinen Wissens* (Leipzig: Verlag des Bibliographischen Instituts, 1890).

## Early Uses of Strongly Electric Fish

It is likely that strongly electric fish first served three purposes for humans: as food, as symbols, and as therapeutic devices. Colonial, missionary, and ethnographic accounts from Africa and South America show that aboriginal peoples were very familiar with strongly electric fish at the time that they first made contact with members of literate societies. For some, the fish served as food, for others, food to be avoided. Strongly electric species had names and played an explanatory role in stories involving gods or godlike transformers. Many cultures used the shock of the fish to treat various ailments. In Abyssinia (now Ethiopia), the shock was apparently used for nervous afflictions. Inhabitants of the Old Calabar River in Nigeria reportedly placed children suffering from fits or colic in a tub of water with a few live electric catfish. This story was repeated in the Victorian press, as the British found it "interesting to find a popular scientific remedy of [their] own, anticipated by the unlettered savage." Despite colonial attitudes, aboriginal people were their contemporaries, not their ancestors, just as present-day hunter-gatherers are our contemporaries rather than a window into our shared stone-age past. The argument here is not that ancient practices were being documented by explorers of various sorts but rather that contemporary practices that are widespread must have originated time and again because they are stable or robust in some way. Strongly electric fish are good to eat, good to think with, and good for what ails you.[33]

Etymological evidence from different times and places suggests that the human experience of shock was typically the most salient characteristic of strongly electric fish. In the Kpelle language of Guinea, the word for the electric catfish is *kplíkplî*, the doubling of the noun suggestive of twitching back and forth. In Mende, spoken nearby, the word is *kpikpi*. In the Hausa language of Nigeria, the same word is used for both the electric catfish and for convulsions, a cramp, or a pins-and-needles sensation. The noun for electric catfish in the Bobangi language of the Upper Congo River is *nina*; the same word used as an intransitive verb means to be cold or wet, while *ninga* means to shake, shiver, or vibrate, and *nga-nga* describes something that is electric or stinging. The Arabic word for the torpedo shares a root with words for a tremor, trembling, shuddering, and shivering, and according to some authors, thunder. In Persian, the word used to describe the torpedo is also used in expressions for fever and quivering. The Ancient Greeks used *narke* interchangeably for both torpedoes and electric catfish; the word also meant numbness or paralysis and is at the root of our "narcosis," "narcotics," and "narcolepsy." The Latin name *torpedo* was bestowed on the electric ray because it protected itself with torpor or numbness. English includes colloquialisms such as "crampfish" and "numbfish" in addition to our Latinate "torpedo." The electric ray is known in Brazilian Portuguese as *tremé-tremé* from the word for shakes. The same expression is used as slang for "a tenement building of ill-repute." The Spanish word *temblador* (trembling, shaking) is used for electric eels and, uncharitably, Quakers (to be sure, that English expression is uncharitable, too).[34]

The oldest pictorial and written depictions of strongly electric fish come from ancient Egypt. People have been selectively and seasonally harvesting fish along the Nile River for the past 40,000 years at least, leaving hundreds of thousands of fish bones at dozens of archaeological sites. Although the vast majority of these remains are from the non-electric catfish genus *Clarias*, we can be sure that the inhabitants of the Nile were just as familiar with *Malapterurus* as their ancestors. An image of *Malapterurus* first appears on both sides of a blue-grey schist tablet called the Narmer (or Na'rmr) palette, dated to about 3,000 BCE. The meaning of this artifact has been widely debated. Some scholars have interpreted the catfish as part of a hieroglyphic rebus spelling the name of a king. Others argue that the catfish was a symbol of the king's control and domination. In elite art of the period, wild animals are often represented as "controlling" other animals, and the catfish would continue to be used as a cultic symbol of power in later Egyptian art.[35]

*Malapterurus* also appears in fishing scenes that decorated the walls of Egyptian mastabas in the form of elaborate reliefs. Mastabas were architectural precursors to the pyramids: rectangular mud-brick tombs with flat roofs and sloping walls. The mastaba of the architect Ty (or Ti) at Saqqara, dated circa 2,445 to 2,421 BCE, contains a number of these scenes. In one of the reliefs, Ty is standing on a reed boat in a papyrus marsh, hunting a hippopotamus. Behind him, one of the boat men has just struck an electric catfish with his barge pole. Under the boat several other fish are represented, all in realistic enough detail that they can be readily identified by genus or species. In another relief, *Malapterurus* appears as one of the fish being caught from a boat with a drag net, and in a third relief it is seen swimming directly under a boat. Both fresh and dried fish were popular foods in ancient Egypt, and fishing with drag nets was common. The nets were long, with wooden floats on the upper edge and leads on the lower. They could be used either from a boat or from shore. As far as we know, *Malapterurus* itself was not mentioned in writing until almost three thousand years later in the *Hieroglyphics* of Horapollo, dating from the fourth century CE. There the fish is referred to as "a man saving others from drowning." The usual interpretation is that the shock of *Malapterurus*, when transmitted through wet nets, caused fishers to release their entire catch.[36]

The strongly electric rays of the Mediterranean also appeared regularly in the sources of classical antiquity, playing familiar roles as food, symbol, and therapeutic device. Archestratus the Syracusan, a poet and bon vivant who traveled widely and remained thin despite sampling every exotic dish that he could find, recommended "A boil'd torpedo done in oil and wine, and fragrant herbs, and some thin grated cheese." Torpedo recipes ranged from the abstemious (athletes were counseled to eat boiled torpedo as part of a balanced diet) to the epicurean (the fish was cooked in elaborate broths and sauces in imperial Rome). Torpedoes also appeared on the menus painted on the walls of the cook shops of Pompeii and were naturalistically rendered in fine mosaic panels that decorated Roman baths, fountains, and pools.[37]

In the Hippocratic and other medical traditions, simples could be made from the tissues of the torpedo, as they were from plants and from the bodies of other animals. Pliny described several such concoctions in his *Natural History*. Alum and torpedo brains mixed together and applied on the sixteenth day of the moon could serve as a depilatory, although it might not have been a very good one, since he recommended that the hair be removed before using it. Another "very marvellous fact" about the torpedo was that "if it is caught at the time that the moon is in Libra, and kept in the open air for three days, it will always facilitate parturition, as often as it is introduced into the apartment of a woman in labour." And for those tormented by lust, "the gall of a live torpedo, applied to the generative organs" could serve as an anaphrodisiac, although one has to imagine that any attempt to procure the gall from a live electric ray might be painful and distracting enough to serve the intended purpose, too.[38]

## Experimenting

Until the twentieth century, the shock of a live electric fish would prove to be more versatile than the flesh of a dead one. In the first century CE, the Roman physician Scribonius Largus began applying live torpedoes to his patients as a remedy for gout and headache. There is no reason to suppose that he was the first person to try this, merely one of the first to write about doing so. In fact, he made a practice of learning remedies from anyone who would share them, "including slaves and wise women." For gout, he advised the patient to stand on the wet ground at the edge of the tide, a torpedo underfoot, until the leg was numb up to the knee. In the case of intractable headache, he gathered two or three live torpedoes and applied them one at a time to his patient's head, removing each when the spot became numb and applying another until the pain stayed away. He cautioned against leaving a torpedo in place for too long because it could permanently take away feeling in the region. For therapeutic use, Scribonius recommended using the "black torpedo," *T. nobiliana*, the largest of the Mediterranean species and the one with the most powerful shock. A generation later, Dioscorides added procidentia of the rectum to the list of maladies that the shock of a torpedo might remedy. These prescriptions were repeated in popular medical literature for many centuries.

As late as 1661, Robert Lovell continued to recommend the shock of the torpedo to "restraine the falling out of the fundaments."[39]

In classical antiquity, the torpedo also became the subject of written natural history for the first time. These were attempts to understand the animal on its own terms rather than in relation to some immediate human need or goal. Whether or not this indicated a new conception of the world is a matter of debate in the history of natural philosophy. G. E. R. Lloyd argued that the Greeks invented the idea of "a *domain of nature* encompassing all natural phenomena," whereas Edward Grant replied that humans have always been immersed in a domain of natural phenomena and "what the Greeks seem to have invented were instructive ways of talking about nature." Either way, beginning around 600 BCE the Greeks were responsible for shifting the balance of explanation from the supernatural toward the natural, and there several depictions of strongly electric fish in this new naturalistic mode.[40]

Fish that were caught to be sold in the markets of the Mediterranean were readily available for study and their characteristics were well known to the people who caught them. This is apparently one of the ways Aristotle learned about the various fish that he discussed in his writings on natural history. He described the anatomy, prey, migratory habits, and reproduction of more than one hundred species, most of them from the oceans surrounding Greece and Western Asia. In several cases, his detailed knowledge of anatomy seems to have resulted from dissection, a skill he may have learned from his father, who was an Asclepiad physician. Aristotle's descriptions of the torpedo are clearly based on both observation of live animals and dissections of dead ones. The torpedo "stupefies any fish it may wish to master, with the peculiar force which it has in its body, and then takes and feeds upon them; it lies concealed in sand and mud, and captures as they swim over it any fish that it can take and stupefy; of this circumstance many persons have been witnesses." He concluded that it "also has plainly caused stupefaction in men."[41]

Other ancient scholars also described the torpedo in naturalistic terms, although there was significant disagreement about the mechanism by which it shocked its victims. In a book on venomous animals, Aristotle's pupil Theophrastus described the way that the torpedo's shock could be transmitted through a trident or a fishing line and rod to the hand of the fisher. Theophrastus also discussed the torpedo in a work on creatures that hide and hibernate, and he may have believed that the fish's shock had something to do with coldness, although this interpretation has been disputed. Diphilus of Laodicea is reported to have made a series of experiments with the fish, determining that only some parts of the live torpedo's body can discharge a shock. Theophrastus and Diphilus served as sources of information for later writers like Pliny and Plutarch, which is how we know about their work. Plutarch wrote that people who found torpedoes that had been stranded alive on land and tried to pour seawater on them, reported feeling "a numbness seizing upon their hands and stupefying their feelings, through the water affected with the quality of the fish." He described the torpedo as "shoot[ing] forth the effluviums of his nature like so many darts" and its prey "being (as it were) held in chains and frozen up."[42]

Routine and extensive experience with strongly electric fish also provided people with new ways of thinking about other phenomena in the physical world. The most notable example comes from an argument about the corpuscular nature of matter in the *Pneumatics* of Hero of Alexandria, although some scholars think that it originated with Strato. Both natural philosophers believed that all matter is shot through with tiny void spaces. Strato, intellectual heir to Aristotle and Theophrastus, used this posited characteristic to explain material phenomena like compression and elasticity. Unfortunately, we only know of his work from later citation, as it did not survive to our time. In a passage in the *Pneumatics*, Hero reasons that void spaces in water allow the rays of the sun to penetrate to the bottom of a container of the liquid. When wine is poured into water, it spreads throughout for the same reason: it is able to penetrate these interstitial spaces in the water. Air must be similar to water because the light rays from multiple lamps pass through one another, illuminating objects from multiple directions in a brightly lit room. Hero goes on to conclude that there must be void spaces in all material, as the shock of the torpedo penetrates bronze, iron, and other solid substances. In this analogy, the shock of the torpedo is like light, traveling through a corpuscular medium to act at a distance.[43]

The fact that the torpedo's electric shock could apparently act at a distance was one of its most notable characteristics, and it begged for explanation. There were a few other natural phenomena that seemed to operate at a distance, such as the attraction of chaff to rubbed amber or of iron particles to a lodestone. Did these operate in the same way or not? Hero of Alexandria understood the action of the torpedo's electric shock by analogy with light. In Pliny the Elder's account, by contrast, such action at a distance could be explained if it were caused by a smell or vapor, and this had therapeutic implications. If there were a power that could affect the human body by "exhalations" or "emanations," Pliny wrote, "what are we not to hope for from the remedial influences which Nature has centred in all animated beings?" When the connection between cause and effect was difficult to determine, the historian Brian Copenhaver notes, ancient theorists like Pliny would make recourse to "*pneumata* or *spiritus*, tenuous material substances or lightly embodied spiritual substances." One consequence of viewing the electric fish's discharge as a vapor was that it might be possible to protect yourself from it by holding your breath, as Engelbert Kaempfer would recommend in the seventeenth century.[44]

The most elaborate model of the action of strongly electric fish in classical antiquity was created by Galen in the second century CE. It was to have a long-lasting influence on European and Islamic thought. Galen distinguished between something like the sting of a scorpion, which involves the direct transmission of a venomous substance, and the shock of a torpedo, which can pass through the trident impaling the fish and into the hand of the fisher. In the Galenic system, the action of the scorpion's sting could be understood in terms of the four elements and their associated qualities. The quality of the animal's venom could affect the balance of humors in the victim directly. But such an explanation could not entirely account for the action of the torpedo's shock. Galen explained the numbness that resulted from being shocked as the transmission of a quality of coldness (much the way that limbs become numb when they are exposed to extreme cold). That the torpedo's action could be transmitted through an intervening substance, however, required an additional, occult factor of some sort. Here Galen had to fall back on analogy with something that seemed to operate in a similar fashion: the magnet.[45]

For Galen, magnetic action involved ethereal spirits that drew iron to the lodestone. He thought the same kind of mechanism explained how a cathartic drug might draw a thorn or arrowhead from deep in a wound and how the quality of coldness might be transmitted from the torpedo through a trident. Galen also used the idea of spirits to explain physiological processes, motion, and sensation. In his framework, the liver created blood from food, allowing natural spirits to flow through the veins to the organs, where they were consumed. Blood depleted of natural spirits returned back via the veins to be replenished in the liver. Natural spirits in the heart mixed with air from the lungs to become vital spirits, carrying heat through the arteries to the rest of the body. Arterial and venous systems were not directly connected: blood depleted of vital spirits flowed back through the arteries to the heart. In the brain, vital spirits were transformed into animal spirits. They flowed from the brain into the nerves, where they displaced other spirits, resulting in muscular contraction and movement. Light united with animal spirits emerging from the eyes, transmitting its quality back to the brain in the form of a spirit that carried a replica of external objects. Later thinkers would keep this basic explanatory mechanism of natural, vital, and animal spirits while changing the interpretation of the spirits themselves from an immaterial or ethereal influence to a material agency, a subtle fluid.[46]

Galen's knowledge of anatomy and physiology came both from the work of earlier physicians and from his own extensive experience. His belief that nerves were hollow—thus allowing animal spirits to flow through their lumina—depended on a knowledge of human dissections and vivisections performed in Hellenistic Alexandria in the third century BCE. There, a rivalry between empiric and dogmatic medical sects set the context for debate about the relative effectiveness of observation versus experiment. Following Hippocrates, the empirics taught that observation of the patient's symptoms was sufficient ground for diagnosis and treatment. In their view, physicians could only learn by observing patients. Dissection provided knowledge of dead bodies, not living ones, and vivisection created abnormal physiological conditions in the process of killing the patient. Neither practice was morally compatible with the vocation of healing. The dogmatists disagreed, drawing inspiration from Aristotle instead. They believed that effective medicine required a deep understanding of anatomy and physiology. Two of the dogmatists, Herophilus and Erasistratus, were allowed by the king to vivisect criminals who had been sentenced to death. Although their own writings did not survive, substantial portions of Herophilus's work *On Anatomy* were quoted by Galen. Herophilus differentiated nerves from other tissue and motor nerves from those that served sensation, noting that both brain and nerve were required to feel pain. He described the brain, eye and liver in great detail and traced the venous and arterial networks. Erasistratus described nerves as hollow, arguing that they carried a fluid to and from the brain.[47]

The performance of human vivisection and dissection was a temporary anomaly in ancient Egyptian and Greek society, as members of both found the mutilation of human bodies living or dead to be repugnant. The Egyptians had a long practice of mummifying humans and other animals, including non-electric catfish, but these activities were sacred and circumscribed. Neither Egyptian nor Greek physicians practiced surgery in anything like our own sense because the lack of reliable anesthetics and of aseptic or antiseptic technique made the prospect appalling and the prognosis unfavorable. Galen served as a physician for gladiators in his youth and would have been intimately familiar with the badly wounded human body. But human vivisection was not permitted to him, and his accounts of human anatomy and physiology are based instead on dissection of other species, particularly the Barbary ape and pig, and on animal vivisection. He was the author of a manual *On Anatomical Procedures*, which would be rediscovered in the sixteenth century and would teach William Harvey and others how to do animal experiments and dissection. In that work, Galen described the effects of fully or partially severing the nerves of living animals, tracing various kinds of paralysis to injuries at different points. His subjects were treated purely as equipment, the cries of the animal serving only as feedback about the function of the affected part. The historian Anita Guerrini suggests that the work of Galen was responsible for making animal experimentation the criterion by which human anatomy and physiology could be understood.[48]

To conclude, we have seen that hominins evolved in environments where they regularly encountered strongly electric fish, and the shocks that these animals produced were their most salient characteristic. People also evolved with a disposition to see the world in terms of equipment which can be repurposed, or taken apart and recombined to new ends. This was (and is) true of organisms as well as inanimate objects. Although it is possible to emphasize the novelty of dissection and vivisection in classical antiquity, these activities are better understood if we situate them in the longer-term contexts of taphonomy, predation, butchery, warfare, animal sacrifice, and haruspicy. Animal bodies (including those of humans) have consistent points of vulnerability, like the jugular vein, and come apart in predictable ways. The development of an array of cutting tools made it easier for people to disassemble bodies and gave the users of those tools a good sense for the normal variation to be found in body parts. Any unexpected deviation from the norm would trigger a search for significance. In other words, a good knowledge of gross anatomy provided an epistemic resource that could be turned to new uses, and this knowledge would have long predated literacy in all human cultures. As an example, G. E. R. Lloyd notes that injuries in the *Iliad* are described in anatomically realistic detail. Treating humans and other animals as subjects for experiment or disassembly lowered the conceptual barriers between ourselves and our animal kin. It also lowered the barrier between animate and inanimate. If an electric fish could be used as apparatus, it might also be possible to build an artificial device that could generate a shock like an electric fish.[49]

TWO

# Modeling Animal Electricity

## Biomimesis

Attempts to make an artificial electric fish eventually led to the human discovery and occupation of an electric world. The practice of drawing inspiration for human-made technologies from nature—called biomimesis or biomimicry—is on the rise in our time, but it is certainly not a new idea. Hominins appropriated the hides and furs of other animals for clothing and shelter. Close study of their natural environments provided them with an ever-expanding range of raw materials and design ideas to put to use. They could ask themselves what characteristics made a predator effective or helped its prey to escape or defend itself. The teeth, claws, horns, barbs, and armor plates of other animals could suggest the shape of a blade, spear, or harpoon or the utility of a handheld shield. Other animals traveled more effectively through water, floated more readily, or moved over the surface of snow rather than being bogged down by it. By creating and using a wide range of artifacts, people could take on some of the desirable properties of other animals at will.[1]

Close observation of animals is one of the most characteristic features of our earliest surviving artworks. In many of these, the desire to assimilate human and nonhuman properties, or the properties of distinct animals, is apparent. The Lion Man of Hohlenstein-Stadel, for example, is a realistic sculpture of a standing man with the head of a lion, carved in mammoth ivory. It was found in a cave in Germany and dated to about 30,000 years ago. A 6,000-year-old bowl from Iran is decorated with a figure that appears to be in a state of transition between bird and human being. The Narmer palette contains depictions of cow-headed goddesses, long-necked cats, bulls, and a hawk, as well as *Malapterurus*. A Mesopotamian sculpture of a muscular lion-headed female figure dates to 2,900 BCE. The Burney relief, or "Queen of the Night," is of a winged and taloned anthropomorphic goddess standing on the back of lions, flanked by owls; from Iraq, it dates to 1,775 BCE. The number of examples could be multiplied almost indefinitely. In fact, it is hard to find instances of paleolithic art that are not zoomorphic in some way, and a great many of these seem to express a desire to appropriate the powers or attributes of other animals.[2]

While exercising a quintessentially human habit of experimenting with other animals, people put strongly electric fish to use in various ways. How was their shock produced? How was it communicated? What effects could it have on the bodies, brains, sensations, or actions of people or other animals? Thinking of the fish as apparatus led to the goal of transferring the animal's power into other artifacts. Strategies for doing so jumped from *semblance* (something that looked like the fish) to *substance* (something made from its body) to *function* (something that acted like it). In some situations, copying the way that something looks is a reliable way to copy its function. The buoyancy of a boat hull, the aerodynamics of a boomerang, or the mechanical advantage provided by an atlatl all follow from getting the form correct. To the extent, say, that snowshoes or skis were inspired by observing animals like the snowshoe hare, the same benefits would accrue from copying form. In other situations the strategy does not work. The ability of the strongly electric fish to generate shocks does not rely on their overall form, which actually varies quite widely. Likewise, it is sometimes possible to extract a scent, venom, drug, or pigment from the body of an animal or a plant, but this did not work for animal electricity.[3]

Nevertheless, most attempts to harness the power of strongly electric fish in classical antiquity relied on similarity of appearance or extraction of substance. This would also be true of the Western, Byzantine, and Islamic inheritors of the classical tradition, and of the contemporaneous African empires and states that were in contact with them. There is one possible exception to this generalization, however, a set of objects known as the Parthian galvanic cells or, more colloquially, the "Baghdad batteries." These were created in the Near East about seventeen centuries before Alessandro Volta demonstrated a device with similar electrical properties. The only hitch is that scholars have not yet come to a consensus about the Parthian cells. They may or may not have been intended as electric batteries, and other explanations have been put forward.

In June 1936, archaeologists from the Iran Antiquities Department working at Khujut Rabou'a near Baghdad discovered an ovoid ceramic jar sealed with an asphalt plug, small enough to be held in one hand. Inside was an iron rod, encased in a soldered copper cylinder. The discovery was first published in 1938 by Wilhelm König, the director of the Iran Antiquities Department and the Baghdad Museum. Since the object had been found in a stratum dating to the Parthian period (first century BCE to first century CE), it was attributed to the Parthians. König searched the archaeological literature and identified some previous finds that appeared to be of the same kind. Small devices stuffed with papyrus and associated with iron and bronze needles turned up in excavations by Leroy Waterman some 40 kilometers away at Seleucia on the Tigris River. In subsequent excavations, Ernst Kühnel found six sealed ceramic jars containing rolls of metal or metal nails. At different sites, wirelike bronze or iron rods were also found associated with the devices. König argued that these objects were electric batteries, and he later suggested that a series of such cells connected together with the wirelike metal rods would provide enough power for gold electroplating.

In order to function as a battery, the iron rod had to be suspended from the asphalt plug so that it did not touch the surrounding copper cylinder, and the space in between filled with a liquid electrolyte of some sort. The asphalt plug would seal the liquid in the cell. In addition, there was a layer of asphalt in the bottom of the device which could hold the iron and copper pieces away from one another preventing accidental contact. Various experimenters have created models of the Parthian cells and tried using several different electrolytes. The main requirement for the electrolyte is that it must conduct electricity and be something that was known to the Parthians. Both citric acid and acetic acid fit the bill, and replicas of the device are capable of generating about one half a volt when filled with vinegar. This is not a very high potential. By comparison, schoolchildren routinely generate higher voltages by sticking zinc nails and copper pennies through the rinds of fresh lemons. Voltage produced by the Parthian cell is almost high enough for gold electroplating, however, but only if one has a supply of gold cyanide. As far as we know, the Parthians did not. There are a few plausible ways that they might have manufactured it, but there is no evidence that they did so. It is also problematic that no electroplated objects were found in the excavations at Khujut Rabou'a. Besides, smiths of the time were already familiar with a range of chemical techniques for metal plating and would not need a less effective one.

Since the mid-1980s, some German scholars have argued that the electroplating hypothesis should be abandoned. They suggest that magicians wrote incantations on organic materials and placed them into the ceramic jars with metals that were thought to have magical properties. In this interpretation, the fact that the Parthian cells can function like batteries is purely accidental. Some of the resistance to the idea of the Parthian cells as batteries comes from a disbelief that there could be any easy way to discover the underlying phenomenon. But touching a pair of dissimilar metals suspended in an electrolyte will result in a tingling sensation. The use of an iron ladle in a copper or bronze bowl of wine or vinegar could easily have led to accidental discovery.

König later provided a different explanation, one that takes the probable electrical function of the Parthian cells into account: that the devices may instead have been used for electrotherapy. At the time, Seleucia was on the trade routes which connected Rome with India and China via the Silk Road. The Greco-Roman practice of therapy using electric rays was known from the work of physicians like Scribonius Largus. Acupuncture was already a standard Chinese medical practice. And Mesopotamian medical tradition emphasized the use of drugs, including many containing vinegar. The classicist (and physicist) Paul Keyser suggests that Parthian medical practitioners brought together these three elements as a substitute for Mediterranean or Egyptian "icthyoelectroanalgesia," since residents of the Tigris-Euphrates region lacked a ready supply of live strongly electric fish. The torpedo generates shocks that are much higher in voltage than any that ancient wet-cell batteries could produce, even when connected in series. But if the current of the Parthian cells was applied subcutaneously, using iron or bronze needles as they are used in acupuncture, then contemporary medical research suggests that local analgesia could be produced with very low voltages.[4]

In *The Shock of the Old*, David Edgerton argues that our understanding of the history of technology has been distorted by putting too much emphasis on novelty and innovation. If we focus instead on "technology-in-use," we find the coexistence of old and new elements. The moment at which something appears to be brand new is often quite distant from the time that it will play the most historically significant role in our lives. And the fact that we are crucially dependent on a particular technique or technology should not blind us to the possibility that it could have been previously explored for other reasons, or abandoned as a dead-end in some other social or cultural context. "Seeing these electric batteries," Paul Keyser writes, "we are struck by the foregone opportunities, but in their context, they were merely one not necessarily very effective tool of practical, magic-using physicians on the edges of the Greco-Roman world." The influence of the Parthian cells did not extend beyond a circumscribed time and place, and electric batteries would have to be reinvented later.[5]

## Natural History

Strongly electric fish served much the same functions for Byzantine, Islamic, and medieval European scholars as they had for those of classical antiquity. Byzantine physicians focused on the therapeutic utility of the torpedo's flesh as food or ointment. In the Islamic medical tradition, by contrast, the shock of the live fish was paramount. Ibn Abbas al-Majusi (Latinized as Haly Abbas) referred to the torpedo as *pisces dormitans*, emphasizing the therapeutic benefits of shock-induced sleep. Ibn Sina (Avicenna) and Ibn Rushd (Averrhoes) recommended the shock for treating headache or melancholy or for stopping an epileptic seizure. Like Galen, Ibn Rushd compared the effects of the fish to the power of the magnet. Although the electric catfish was routinely confused with the torpedo until at least the eighteenth century, the twelfth-century Islamic physician and naturalist Abd al Latif gave a good description of

*Malapterurus* in his history of Egypt. "While it lives no one can touch it without experiencing an irresistible trembling sensation," he wrote. "This impression is accompanied by cold, numbness, a crawling feeling and lameness in the limbs, so that it is impossible for one to remain in an upright position or hold anything." The slightest contact with the fish could be felt through the arm, shoulder, and side, and bathers warned al Latif that the "mere breath" of the fish could affect a person to the point where he or she was in danger of sinking into the water. In Europe, Albert the Great recorded the account of a man who brushed a torpedo with his fingertips: "right away his hand and arm up to the shoulder became so numb that with much use of hot baths, poultices, ointments and massage he could scarcely recover feeling and movement in his arm in less than a month." The man was still suffering after half a year.[6]

The electric catfish also played an important symbolic role in the west African kingdom of Benin, which flourished in what is now southern Nigeria between the twelfth and late nineteenth centuries CE. Animals were frequently depicted in the sculpture of Benin, where they were used to represent the overlapping boundaries of the human, nonhuman, and supernatural. The "mudfish" had a prominent part in this iconography. The attributes of different African catfish species were combined into a chimera or composite that could variously discharge electric shocks, sting with venomous spines, and breathe and "walk" on land. In its latter capacity it served to exemplify utility to humankind. "The mudfish is the freshest, most robust and most delicious of all fish and is considered very attractive and desirable," the anthropologist Paula Ben-Amos wrote. "It represents prosperity, peace, well-being and fertility through its association with the water, the realm of the sea god, Olokun." Some catfish aestivate in the mud of dry streams, breathing air until the rains seem to bring them back to life. This aspect of the mudfish made it a symbol of power and transformation. Sculptures of Benin kings show them standing on mudfish or show their legs transforming into mudfish. In addition, the electric shock and poisonous stings of the mudfish represented its danger and authority. Brass masks depict kings with beards made from stylized mudfish, each projecting from the face in the same way as the barbels of a catfish. The mudfish's representational liminality—between water and land, prosperity and danger, human and nonhuman—was also used to characterize the Portuguese after their arrival in 1486. An armlet made from ivory with inlaid brass shows European heads alternating with those of mudfish.[7]

Discussions about the cause of the torpedo's action flourished in sixteenth-century Europe. Initially these were quite similar to those that came before, as Renaissance natural philosophers looked to antiquity for inspiration and as the new technology of printing gave them a level of access to ancient texts that would have been beyond the wildest dreams of medieval scholars. Plato's remarks were perhaps best known: in the dialogues, Meno accused Socrates of being like a torpedo who caused his soul and tongue to become torpid. Erasmus quoted this passage in his *Adages* and compared evil speech to the action of the torpedo's poison. Poison was not the most reliable explanation for the torpedo's powers, but it made sense in a way. Paresthesia that is caused by the consumption of poisonous fish like puffers results in a tingling, electric-shock feeling in the extremities. Even if few people were unlucky enough to experience both phenomena (or lucky enough to do so and survive), the reports of the victims of each misfortune would have had much in common. The flesh of the torpedo could be eaten without ill effect, however, so poison remained a problematic explanation.[8]

Over the course of the sixteenth century, the authority of ancient texts was undermined by a continual stream of new information, not all of it veridical, flowing through rapidly expanding European networks of exploration, trade, and exploitation. New claims could still be made about the familiar electric ray, and reports of other fish with similar properties began to circulate. In 1556, Jean de Léry, the French Calvinist minister at a Rio de Janeiro Huguenot colony, reported that he saw a ray sting the leg of a man as they were pulling it into the boat, leaving the spot red and swollen. From his description of the injury and the fish, this was most likely a poisonous stingray rather than an electric ray, but that distinction may not have been clear to people reading it at the time.[9]

There was much more to learn about the familiar torpedo, too. While working for the Dutch East India Company, the German physician Engelbert Kaempfer made detailed observations and dissections of a new species of torpedo in the Persian Gulf. When captured or handled, the fish voluntarily struck "with a sort of momentary belching or a certain convulsive motion of the viscera, whereby it dilates the spiracles of the abdomen and absorbs air; with the same effort it simultaneously thrusts out its dreadful virus into the air." Kaempfer compared the fish's reaction with the porcupine's ability to "eject spines at those annoying them" (alas, we no longer believe that this can happen). The strike was less powerful when the fish was handled in water, less powerful if handled for a number of hours and did not travel through an intervening staff or spear. An African man showed Kaempfer that he could hold the fish without injury if he held his breath at the same time. Kaempfer thought the process of forcefully exhaling before drawing the breath to be held was responsible for dispelling the "miasma" of the fish. He also described the experience of shock carefully:

> The numbness induced is not the sort felt in a sleeping limb, but a sudden condition that instantly travels through the touching part and penetrates the citadel of life and breath. Then it overwhelms the whole body and mind, as it seizes the sinewy and bony parts such as the hands, shins, and elbows. In a word, you would think that your major joints were broken and limp, especially those of that member which first received the expelled vapors. And all of this is accompanied by a shudder of the heart, a trembling of the limbs, a numbness, and a chill. So powerful and so swift is the force of the horrifying exhalation that like a chill bolt of lightning it shoots through the handler.[10]

Seventeenth-century reports from Africa suggested to European scholars that the strongly electric fish found in the Nile and other rivers there might not be a ray after all. In 1597, a Portuguese missionary reported on the *tremedor* or *thinta*, a strange fish common in the Sofala River of Mozambique. Once dead it was prized for its meat, but no one could touch it while alive "for it filleth the hand and arme with paine, as if every joynt would goe asunder." He was told that its skin was "used to sorceries" and that the fish could be roasted, ground to powder, and served in wine as a treatment for colic. Traveling up the River Gamba into the West African interior in 1620–21, the Englishman Richard Jobson described taking a similar fish in a net and discovering that it shocked the hands and legs of his sailors. "Their sense being seene to returne, the Cooke was called up, and bidden to dresse it; who laying both his hands thereon, sunke presently on his hinder parts, making grievous moane that he felt not his hands." An African man in a nearby canoe laughed and told them that the fish was feared while alive but good to eat when dead.

Using strongly electric fish to shock unsuspecting victims as a crude joke would become a relatively common anecdote in European travel literature. Portuguese missionaries in Ethiopia described the strongly electric catfish in their letters and manuscripts. In 1615 the Jesuit Nicolas Godinho (or Godigno) reported that the fish was used by the Ethiopians to "cast out devils." Although he never witnessed it, they also told him that if a living electric catfish was tossed on a heap of dead fish "the fish thus brought in contact with it are seized with an inward and inexplicable trembling to such an extent that they actually appear to be alive. The cause may be authenticated by those who investigate the nature of things in general, and I leave it to them to decide as to what this force of motion communicated by the electric fish to the dead ones may be."[11]

By the 1670s, news of a third strongly electric fish was circulating in Europe, this time from South America. The fish was actually first mentioned in Fernández de Oviedo's *Natural and General History of the New World* (1526), where it was described as a torpedo, capable of communicating a shock through a wooden spear. Sixteenth-century Portuguese Jesuits also described the *poraquê*, or electric eel, albeit in torpedo-like terms which made it difficult to know that a completely different kind of fish was being depicted. In retrospect, we can find the strongly electric eel mentioned again and again in European writing, confused with the torpedo for more than a hundred years. By the late seventeenth century, however, it was clear to thinkers in Europe that the electric eel was different from both the electric ray and the electric catfish.

In the 1670s, the astronomer Jean Richer made an expedition to Cayenne (in what is now French Guiana) where he observed an eellike fish that was three to four feet long, "as fat as a leg" and capable of numbing his arm for 15 minutes when touched with a finger or the tip of a stick. The English writer Aphra Behn encountered the electric eel on a visit to Surinam in the 1660s and recorded eating one. The fish played a more dramatic role in *Oroonoko*, Behn's novel of Surinam, where it almost kills her protagonist. Revived after much ado, he recovers enough by the evening to dine on the flesh of the fish. The electric eel was also of interest to the savants of the Royal Society, founded two decades earlier. On March 18, 1680, John Evelyn noted in his diary that there was a letter from Surinam that described a small eel, that when hooked at distance of 100 feet "did so benumb, & stupifie the limbs of the Fisher, that had not the line suddainly beene cutt, by one of the Iland (who was acquainted with its effects) the poore man had immediately died: There is a certaine wood growing in the Country, which put into a Waire or Eele-pot, dos as much intoxicate the fish as Nux Vomica dos other fish, by which this mortiferous Torpedo is not onely caught, but becomes both harmlesse, & excellent meate."[12]

## Natural Magic

Prior to widespread acceptance of mechanistic thinking in Europe, strongly electric fishes were one of a number of phenomena that were used to justify natural magic. Galen had reduced some of the fish's ability to the perceptible quality of coldness, but its action at a distance required him to posit an occult factor. Other immaterial explanations followed in this vein, attributing the shock to spirits, smell, vapor, aura, or sympathy. In 1580, Michel de Montaigne suggested that the torpedo must be conscious of its own "miraculous" ability and perhaps even use it to sense its prey while it hid itself in the mud, an intriguing suggestion that was not followed up. Even imagining the shock as a kind of poison still entailed something occult, to distinguish the fish's abilities from the more straightforward ones of a venomous snake, tarantula, or scorpion. As such, strongly electric fish were placed in the same category as magnets, amber, or the moon, which had analogous abilities to act at a distance, drawing iron particles, chaff, and the tides respectively. The mid-sixteenth century was a tipping point between "the poverty of erudition," represented by works that depicted the torpedo in essentially ancient terms, and the new ichthyological studies of people like Guillaume Rondelet (1554–55), which were based in part on personal observation. Trained as a physician and anatomist, Rondelet compared the sensation of touching the fish to the effects of opium, mandrake, and henbane. In the same year, the papal physician Ippolito Salviani reasoned that since dead animals become colder and since a dead torpedo loses its power, the power of the torpedo cannot be due to the quality of coldness.[13]

Occult qualities were central to traditional physics, metaphysics and medicine, and their gradual elimination was the work of a few generations. Over the course of the seventeenth century, ancient atomic and corpuscular theories were revived to take the place of explanation by occult qualities. In this view, causation could only follow from motion, and motion from direct contact. So infinitesimal particles were posited, thought to be emitted by one body, travel some distance, and then enter the close-fitting pores of another, mechanically affecting it. These subtle particles could take the form of invisible hooks, rods, strings, or screws, thus providing a mechanical explanation for the electric ray's "stupefactive emanation." Such mechanical accounts were not entirely convincing, even to those generally sympathetic to mechanistic explanation. In the late eighteenth century, Robert Boyle would persist in describing the power of the torpedo as a poisonous exhalation. But whatever the difficulties with corpuscular explanations, subsequent accounts would increasingly share an emphasis on mechanism.[14]

In Florence, the Medici Granducal court supplied torpedoes to members of the Accademia del Cimento and a few other natural philosophers and natural historians. Francesco Redi experimented with and then dissected a torpedo in March 1666, writing up his findings in a letter to the Jesuit polymath Athanasius Kircher. Redi provided the first description of the fish's electric organs as such (or at least the first to survive in the written record). He noted that the painful sensations caused by the fish were most intense when these sickle-shaped bodies were touched. Redi believed that the shock must be due to mechanical action but did not explain how he thought this worked. His correspondent Kircher preferred an occult explanation, going so far as to claim that "magnetic virtue" could pass from the torpedo's stomach to other organisms that it had swallowed. Redi and Kircher subsequently debated the possibility that a stone that supposedly originated in the head of Indian cobras could sympathetically suck poison from wounds. Redi thought not. Redi's pupil Stefano Lorenzini also dissected the torpedo, identifying nervous structures that are now named for him and speculating that the muscular contractions of the electric organs must result in the emission of corpuscles that "fitted to the pores of a man's skin, so as to enter upon immediate contact, but not otherwise, . . . disturb the posture of the parts, and . . . cause pain, as when one's elbow is hit or knocked." Lorenzini rejected the possibility that a live electric fish could affect newly dead ones, as in the Jesuit missionary Godinho's report.[15]

The occultism of people like Kircher did not sit well with members of the Accademia del Cimento, who were inspired by Galilean ideals. They also rejected Galenic animal spirits. In a demonstration, Giovanni Borelli submerged the limb of a living animal in water and then cut into its muscle, showing that no spirits came bubbling out. Borelli was a mathematician and an advocate of iatromechanics, an attempt to create a systematic medical theory based on the principles of mechanics. He was also a proponent of biomimicry, arguing that a mechanical analysis of the swim bladder of a fish suggested that it was possible to create underwater breathing apparatus and submersible boats. While experimenting with strongly electric rays, Borelli showed that certain parts of the torpedo's body could be touched without drawing shocks, and that it could sometimes be handled without effect while it rested. This strongly supported the hypothesis that the shock should not be attributed to an occult quality of the animal as a whole but rather to some local mechanism it possessed. Since the feeling of being shocked was similar to that of a blow on the elbow, Borelli imagined that the effect was due to the person's fingers being struck repeatedly by rapid muscular motions of the fish. Despite the fact that Borelli's account of the torpedo's shock could not explain how it was transmitted through water, his proposed mechanism was accepted as the best one available for almost a century.[16]

## The Leyden Jar

Both mechanical thinking and the quantification of experience changed the human relationship with strongly electric fish, and this change would eventually have far-reaching consequences. As we have seen, the ability to view the world in terms of mechanisms that could be disassembled, understood, and adapted to new uses is a trait with deep roots in the hominin lineage. Progressive and thoroughgoing quantification, by contrast, emerged in western Europe in the Middle Ages. There, many anonymous craftsmen experimented extensively with mechanical devices in order to put them to use in productive processes. Although many of these mechanisms were well known to scholars in classical antiquity like Archimedes and Hero, in the ancient context they were little more than curiosities. During the medieval expansion of European cities and the concomitant rise of merchant and professional classes, however, an emphasis was put on utilitarian and practical values. A new dialectic emerged: it could be rewarding to think in terms of applied mechanics, and the multiplication of machines created niches for people who could do so, leading to the further elaboration of a mechanical world.

The most precise of these machines allowed space, time, and experience to be quantified. The creation of the verge and foliot escapement in the second half of the thirteenth century provided mechanical clocks with a steady oscillator that could divide time into uniform quanta. Before this, the best clocks relied on the regulation of a continuously varying quantity such as a steady flow of water or sand or the fall of a weight. Time divided by a mechanical clock more readily lent itself to counting and measurement, and the impulse and means to do so spread to other domains. In other words, it was now possible to represent other measurements per unit of time. Space was quantified, numerical manipulation became more sophisticated, and music, painting, and bookkeeping became domains of measurement and notation. Subjective experience, too, would eventually become something that could be calibrated with external measurements.[17]

The word *electricity* was coined in 1600 by the London physician William Gilbert, in his work *On the Magnet*. Few prior natural philosophers had distinguished magnetic attraction from that between chaff and rubbed amber. Gilbert proposed to describe the latter as "electric" attraction and noted that glass, gems, ebonite, resin, sealing wax, and other materials behaved the same way as amber when rubbed with silk or wool. He believed that when these electric materials are rubbed, they release natural emanations that cause attraction. Gilbert's method was resolutely experimental. He made use of a device that he called a "versorium" (subsequently known as an electroscope), a suspended metal needle that would revolve in the presence of rubbed electrics but not magnets. Using the instrument, Gilbert determined that electric attraction was subtle enough to act through a coating of olive oil but could be disrupted by intervening substances like linen, paper, or brandy. He also compared the sparks of rubbed electrics with those thrown off by flint: only the latter were capable of causing fire. By contrast, the lodestone did not create sparks but could act through every intervening substance that he tested, including marble. Unlike the mechanical mechanism that he proposed for electrics, Gilbert's explanation for magnetic attraction was framed in terms of sympathies.

The distinction of electricity from magnetism altered the space of possible analogies for the action of strongly electric fish, both making a mechanical explanation seem more likely and providing the apparatus and techniques necessary to refine the understanding of that mechanism. As long as the fish was understood in analogous terms to a lodestone, it was susceptible to explanation in occult terms. But, as Heilbron argues, even seventeenth-century scholars who explained magnetism by reference to occult qualities (like the Jesuit polymaths or Gilbert himself) preferred a mechanical explanation for the action of electrics. The reason for this was that electric attraction was exhibited by a such a diverse range of substances that it did not seem possible that it could be due to some occult quality that they shared. Gilbert's electroscope provided a convenient way to test for the presence of electric attraction.[18]

In the second half of the seventeenth century, experimenters developed machines to generate electric attraction by friction. These involved spinning a sulphur or glass globe rapidly against a piece of cloth or the experimenter's dry hand. The spinning globe would get warm and begin to crackle. It also attracted light objects like feathers, chaff, or bits of paper, which clung to the surface as it spun. In a darkened room, hands could be seen to glow with a purple light when they touched the globe. By the early eighteenth century, electrical demonstrations and experiments were being performed at the Royal Society and other places. But electricity could only be generated in small bursts and discharged more or less immediately. As a result, versions of friction machines circulated widely and were subject to continuous refinement. Various kinds of mechanism were added to allow the globe to be turned more easily or rapidly, spring-mounted leather pads took the place of a dry hand, metal combs and rods conducted electricity from the charged globe to be transmitted to a more convenient position near the device. Some machines even had primitive electrical detectors of a sort: charged woolen threads that pointed toward or away from electrified objects.[19]

A self-taught, provincial artisan named Stephen Gray demonstrated that it was possible to communicate electricity over ever-increasing distances. He started by showing that a down feather was attracted to a glass tube once the tube had been rubbed, but then it might oscillate between the tube and another nearby solid body, or hover, or move back and forth in time to the motion of rubbing on the glass. Gray's conclusion was that currents of effluvia might be emitted and received by all objects. If so, this current flow was considerably more complicated than had been suggested by earlier experiments at the Royal Society. Gray spent a number of years trying unsuccessfully to electrify metals with friction, heat, and percussion. In 1729, he hit on the idea that since a glass tube could communicate light under some conditions, it might also be able to communicate electricity. While doing tests with a flint glass tube stoppered with a cork, he noticed the feather was attracted to the cork rather than to the rubbed tube. He then mounted an ivory ball on a stick and drove the other end into the stopper. The ball attracted the feather even more than the cork itself had when the glass tube was rubbed. He tried varying the connection between the cork and ball, using thread and metal wires, and substituting other objects for the ball, such as coins, stones, vegetables and a tea kettle. When the flint glass tube was rubbed, any metals connected to it became strongly electrical. In subsequent trials, Gray and his associates managed to increase the distance by which electricity could be communicated to over 50 feet. They also demonstrated that a wide range of materials could conduct electricity, if a charged tube was touched to one end of the conductor and bits of paper, down feathers or a piece of leaf-brass were placed near the other end. His most compelling conductor was a human being, a 47-pound charity boy suspended in midair by cloth lines.[20]

The wide range of electrical phenomena exhibited in the 1730s was difficult to square with any version of effluvial theory. In 1733, Charles François de Cisternay Dufay reviewed the previous work on electricity and then set out to bring some order to the observations. He argued that all solid materials except metals can be electrified by friction, eliminating a hodgepodge of "electrics" in favor of a nearly universal property of matter. He introduced the idea that a body must be resting upon a thick electric support in order to become electrified—hence the suspension of the electrified boy, who would not have become so had he been standing on the ground. Besides air, sheets of glass or wax also provided good platforms for bodies that one wished to electrify, whereas wood or metal supports did not. Even water could be electrified if its vessel were placed on a wax or glass support first. Dufay also showed systematically that there were three phenomena that had to be accounted for by a theory of electricity: attraction, conduction, and repulsion. To explain these, he eventually settled on the hypothesis that there must be two kinds of electrification. Bodies with opposite values would attract one another, while those with like values would be repelled.[21]

The inability to store electric charge, or to dispense it in a regulated manner, remained a hindrance to experimentation. In fact, all of the raw materials needed to create a condenser to store and concentrate charge were at hand: the friction machine, wire, and glass vessels of water. Prevailing theory, however, made it unlikely that the elements would be correctly combined by intention. Instead the device had to be accidentally and independently discovered in Germany and in Holland in the mid-1740s, by people who were not familiar with the way that electrical equipment was supposed to be operated. The dean of a cathedral chapter in Pomerania, Ewald Jürgen von Kleist, tried running the electrical charge from his homemade friction machine into a glass of water. Dufay had shown that bodies to be electrified needed to be insulated, so Kleist did so, electrifying the water as others had done. He tried electrifying a wooden spool with a metal nail driven into it, then hit on the happy idea of placing a nail into a small glass medicine bottle of alcohol before electrifying it. Once he disconnected this phial from the friction machine, he could carry it around the room discharging sparks. Kleist also discovered that if he tried to touch the nail in the phial while he was charging it, he received a strong shock. If the glass were partly filled with water, the results were even stronger. When he built a similar device with a wire running into a thermometer tube filled with alcohol, "according to the kindly dean, children of eight or nine could be knocked off their feet by the shock." Although he described the device to correspondents, none could replicate his effects. This turned out to be because he was holding the outside of the phial while he charged it, contrary to standard practice, which would have been to insulate the phial instead. But Kleist's phial would not work if it were insulated while charging.[22]

Pieter van Musschenbroek, a professor of experimental physics in Leiden, was also experimenting with electrifying water around the same time. He knew enough not to touch a glass vessel while it was charging, lest the electricity be lost by connection to ground. But a lawyer friend, while entertaining himself alone in Musschenbroek's lab, did try holding a glass jar of water while charging it and received a painful shock. He reported the occurrence to Musschenbroek, whose assistant subsequently managed to knock the wind out of himself in the same manner. A few days later Musschenbroek tried the experiment himself and was badly hurt and frightened. His letter to colleague René-Antoine de Réaumur at the Paris Academy subsequently became quite well known. It began, "I would like to tell you of a new but terrible experiment, which I advise you never to attempt yourself, nor would I, who have experienced it and survived by the grace of God, do it again for all the kingdom of France." He went on to say, "my right hand was struck with such force that my whole body quivered just like someone hit by lightning . . . the arm and the entire body are affected so terribly I can't describe it. I thought I was done for." Despite his dire warning—or more likely because of it—many people would successfully replicate his experiment, refining the equipment in the process.[23]

By the mid-1740s, then, natural philosophers and natural historians had an artifact that could be articulated both with electrical phenomena and with longstanding ideas of "animal spirits" or "vital fluids." The improved version of the Leyden jar consisted of a glass vessel partially sheathed in metal and capped with an insulating plug. A metal lead connected the interior of the jar, which was partially filled with water or metal shot, through the plug to a ball or ring on the outside top. The Leyden jar was charged by connecting the outer surface of the jar to the ground while touching the metal ball to the conductor of a running friction machine. It was discharged when a conductor—animal, vegetable, or mineral—touched both the ball and the outer sheath at the same time. Experimenters soon discovered that the shape of the jar didn't matter; that it could hold a charge for hours or days; that jars could be connected to one another to form a "battery" and that "a stroke from two such phials, taken to the forehead, felt like a blow from a bludgeon." Jean-Antoine Nollet, physician and priest, demonstrated that he could make hundreds of gendarmes or monks jump from the shock if they held hands in such a way as to make a circuit that closed the ends of a charged battery. Extending nonhuman circuits by a mile or more, experimenters could still not perceive a delay between seeing the spark and feeling the shock. Electricity seemed to travel instantaneously.[24]

By the early eighteenth century, the best explanations for the action of the torpedo and other strongly electric fish were mechanical, but they still failed to account for the animals' ability to act through water and at a distance. In 1714, René-Antoine Ferchault de Réaumur observed torpedoes caught on the French coast near La Rochelle. In one of his experiments, he put a duck and a torpedo in a container of sea water, covering the top with linen so that the duck could breathe but not escape. After a few hours the torpedo had killed the bird. Réaumur observed that the torpedo's back flexes upward to become convex when it discharges a shock, an effect he compared to the action of tiny mechanical springs. He thus accepted Borelli's explanation that the shock was due to a series of sharp blows caused by rapid muscular contractions of the fish and rejected ancient claims that the shock could travel through water or intervening bodies. Although some scholars seemed willing to treat the matter as settled, within the span of a few decades the action of strongly electric fish would come to be attributed to just that: electricity.[25]

Friction electric machines, Leyden jars, and electroscopes could be transported easily from one place to another in the eighteenth century. Strongly electric fish could not. In Surinam in 1745, the English surgeon Dale Ingram had the chance to experiment with an electric eel that had been brought into the city alive in a shallow tub by some aboriginal people. Upon bringing his finger to within an inch of the fish, Ingram wrote, "as quick as lightning, my elbow received such a strong repelling force accompanied by such anguish that I thought my fore arm would have fallen off." Taking an iron hoop from an old Madeira barrel and straightening it, he attempted to touch the creature, but the rod was knocked from his hand as if he had been disarmed by an opponent while fencing. He was told that the Indians were able to seize the animals by the back in such a way as to "defeat all their electrical energy or spring." Ingram's letter, published in 1750, was subsequently translated into German and reprinted a number of times.[26]

Stuck in England, the natural philosopher Robert Turner also believed in the possibility of piscine electricity, but he lacked firsthand experience with it. In his 1746 *Electricology*, he argued that the torpedo is "electrified by Nature . . . as the ether is continually pervading the Pores of all Bodies, it may from their particular Textures and Juices be accumulated and retained in them (like as it is in the electrified Phial) till some Body be brought near them, at which Time the Virtue will be dicharg'd upon it." He noted that cats, horses, and some men had been found to be electric, so why not fish or even vegetables? Lacking an actual torpedo with which to experiment, Turner improvised. He laid a flounder on a tin plate and electrified it with an electrical machine—"it becomes an artificial Torpedo, and acts in every Respect, as a Natural one does." Turner's electrified flounder could deliver a shock that was felt from fingertips to elbow, and struck through thin cloth or an iron bar but not a soft stick. Although Turner believed that he had resolved the question of the torpedo's power once and for all, his work does not seem to have made much of an impact on his contemporaries.[27]

A similar suggestion that strongly electric fish might themselves work in the same manner as a Leyden jar was made in 1751 by the French botanist Michel Adanson, who encountered the electric catfish *Malapterurus* on an expedition in Senegal. He wrote that when touched directly or with an iron rod, the shock of the catfish was not "sensibly different from the electric commotion of the Leyden experiment" that he had performed a number of times. Adanson's natural history of Senegal was published in French in 1757 and translated into English, but it did not have seem to have an effect on English-language research any more than Robert Turner's work had. We can see, however, that in the Leyden jar natural philosophers had finally found an artifact that could faithfully replicate a particular experience. The shock of a strongly electric fish might be similar to coldness, to poisoning, to a pins-and-needles sensation, or to a blow on the elbow, but it was the same as a shock from the electric phial. An experience that differed in degree but not in kind could eventually be measured because apples were no longer being compared to oranges. The historian Jessica Riskin argues that the 1750s marked a turning point in French natural philosophy, as the idea of "sensibility" came to represent both physical sensation and moral sentiment serving as the foundation of all knowledge. In this context, it made sense for Adanson's empirical knowledge to originate in feeling, in both senses of the word.[28]

In the 1750s and 1760s, most experimentation with strongly electric fish was done in the Dutch colonies of South America. Following the invention of the Leyden jar in Holland, Dutch physicists began a program of extensive experimentation with electricity. Furthermore, Dutch collectors and scientists were wealthy enough to indulge their interests in natural history, acquiring exotic specimens such as the electric eels that were plentiful in their colonies in the Guianas. The ease with which a Leyden jar could be constructed, charged, and transported made it what Bruno Latour refers to as an "immutable mobile." It was a reliable and convenient means of moving a phenomenon from European centers (such as a laboratory in Leiden) to the more out-of-the-way places that strongly electric fish were typically encountered (like the rivers of the Essequibo colony in Dutch Guiana). The Leyden jar allowed Europeans to act at a distance, constructing knowledge that could be returned from the periphery in the form of other immutable mobiles, typically written records of experience.[29]

Jean Nicolas Sébastien Allamand, a physicist and colleague of Musschenbroek in Leiden, was also a natural historian and collector. Familiar with Richer and Ingram's reports on the electric eels in French Guiana and Surinam respectively, Allamand wrote to Laurens Storm van 's Gravesande, the governor of the Essequibo colony, to see if he could learn more about the electric eel and obtain a preserved specimen for his cabinet. His reply (in French) was translated into Dutch by Allamand and published with a commentary in 1756. In the letter, 's Gravesande confirmed that the sensation of touching the electric eel was the same as that of touching a charged Leyden jar, save that the fish did not emit sparks or fire. "But for everything else it is the same, yes, even much stronger, because if the fish is big and lively, the shock produced by the animal will throw anyone who touches it to the ground . . . All this happens in an instant." 'S Gravesande offered to send living specimens of the electric eel back to Europe, saying that they needed only small crabs to eat. Allemand added a further plea for live specimens in the published version of the letter. He promised a rich refund to anyone who had properties in South America and was willing ship electric eels back to Europe. If the conditions were right, animals themselves might serve as immutable mobiles, but that was difficult to accomplish, as it required the re-creation of the animal's niche. Living electric eels would not be transported successfully to North America or Europe for almost two decades.[30]

Two years later, the collector Laurens Theodore Gronov published a report on the electric eel. He had a dead specimen that he was unwilling to dissect, which he described in detail. He had also asked his correspondents in South America to answer twenty-five questions, many about the conditions under which the creature's shock would be felt. For example, could you feel it if your hand were covered with oil or wax? Yes. Did it feel the same if you touched it while it was in a wooden or a stone container? Yes. Although Gronov's report did not do much to advance understanding of fish electricity, it did serve as a reference for Musschenbroek, whose introduction to natural philosophy was published posthumously in Latin in 1762. Unlike earlier Dutch scholars—who had been ambivalent about a range of potential mechanisms including Borelli and Réaumur's mechanical hypothesis—Musschenbroek was willing to claim that both the electric eel and the torpedo generated electricity. The Dutchman Frans Van der Lott, who was stationed in Essequibo, concurred. In a letter dated June 7, 1761, and published the following year, Van der Lott argued that the power of the fish was the same as that generated by electrical apparatus, save that the fish did not generate sparks. This observation was to become a crucial point in subsequent debate.[31]

The Dutch colonists performed a range of experiments with the electric eel. Van der Lott tried touching the fish with various metals, like copper, tin, gold, silver, and lead, discovering that a more violent shock resulted from some metals than from others. The strength of the shock also varied with the location that the animal was touched, being most intense directly under its throat. At 's Gravesande's house, five persons in a row held hands and one at the end touched an electric eel with a sword. Although all could feel the shock, it was weakest at the end of the chain. Adriaan Spoor had an electric eel placed in a twenty-six-foot-long trough made from a hollow tree. When Van der Lott touched the creature at one end of the trough, Spoor, whose hand was in the water at the other end received a strong shock. Reports of these activities and others soon spread through European scientific, medical and popular literature, most significantly in German language publications. Linnaeus drew on the Gronov's report for the twelfth edition of his *Systema naturae*, giving the fish the name *Gymnotus electricus*. The French were more reluctant to abandon the mechanical hypothesis in favor of an electrical one. The surgeon Philippe Fermin, in Surinam from 1754 to 1763, dissected an electric eel, drawing attention to muscles in its back and breast that he thought were responsible for its ability to shock, by providing a short, sharp jerk. Finding fault with the arguments of Musschenbroek, the secretary of the Académie Royale de Sciences Grandjean de Fouchy went so far as to suggest that "since there are two thousand leagues from [Paris] to Surinam, the facts can be strongly altered during the journey." There was apparently still some mutability left in the mobiles.[32]

Most knowledge of electric eels in English-language research originated with Edward Bancroft's *Essay on the Natural History of Guiana* (1769). Bancroft knew Van der Lott and cited his work. In the essay, Bancroft described a freshwater fish that he provisionally called the "Torporific Eel." Touched by a metal rod or with the bare hand, or (he was told) "by a stick of some particular kinds of heavy American wood," the fish discharged a violent shock that exactly resembled an electric one. It was painful enough that few people wished to repeat the experience. Bancroft couldn't confirm the claim that the fish's shock was conducted by wood, as he failed to receive a shock when he touched the swimming fish with sticks made from oak, ash, or other kinds of wood. Although Bancroft didn't have any experience with the torpedo, he surmised that the shock of both fishes was caused by the same kind of mechanism and transmitted the same way. Bancroft noted that a person who caught the fish on a line immediately received a shock; that a strong shock could be communicated through eight or ten feet of water; that—exactly as with an electric machine—ten or twelve people holding hands would receive a violent shock if one on the end touched the eel with an iron rod; and that a person could unexpectedly feel a shock when they held their hand in the air five or six inches above an enraged fish. He therefore concluded that the fish's shock was due to "an emission of torporific, or electric particles."[33]

## Calibrating Experience

Having established that the experience of a shock discharged by an electric fish was qualitatively the same as that from a Leyden jar, natural philosophers were in a position to try to quantify electrical phenomena. Bancroft's report stimulated a new round of experimentation with torpedoes in Europe. With the encouragement of Benjamin Franklin, John Walsh went to La Rochelle and the Isle of Re in France in 1772 to study the torpedo. Writing back to Franklin in July, he reported that the fish formed circuits through conductors like water and metal and was interrupted by nonconductors like glass and wax; thus, it seemed to be electrical in nature. As in the experiments of the Dutch colonists and Edward Bancroft, the bodies of Walsh and his colleagues played an essential role, for they formed a part of the circuits that he explored. Standing on an insulator of some sort and holding hands in a chain, four people felt a weak shock when the two at the ends used their free hands to touch the fish. Two people could get the same result if they held the opposite ends of a wire while they touched the fish. As they shocked themselves, they estimated the current by their sensations: did they feel it only in their fingers, or did it extend "above the elbow"? When a piece of glass was substituted for wire in the circuit, they felt nothing. The experimenters were also more than willing to put other people's bodies into their circuits. Inspired by Walsh, Jan Ingenhousz used a Leyden jar to shock some fishermen, who assured him that the sensation was the same as the one that they experienced when they touched a torpedo.[34]

No one had yet been able to get the torpedo to create a visible spark or an audible snap. Writing to Franklin later from Paris, John Walsh described his attempts to produce a spark. Working both during daylight and in complete darkness, he tried to create circuits with a narrow enough gap that a spark could jump across. In one arrangement Walsh connected the torpedo to the ends of a chain suspended freely, but he could not see sparking in the minute spaces between the chain links. He also tried running leads from the fish to a piece of tinfoil pasted on sealing wax, but after he had cut "an almost invisible separation" in the foil with a penknife, he still could not see a spark. Despite his efforts, he was unable to "force the torpedinal fluid across the minutest tract of air." But Walsh felt that his experiments were only the beginning. Artificial electricity had served to elucidate the workings of the torpedo, and the study of animal electricity, in turn, might shed light on artificial electricity. The apparent differences of the two kinds of force seemed an obvious starting point for investigation. Within the next few years, Walsh obtained a much more powerful *Gymnotus* specimen, and in the summer of 1775 he used the electric eel to demonstrate a perceptible spark through the air. For many natural philosophers, the spark was necessary to prove that the power of strongly electric fish was electrical in nature. "He, who analyzed the electrified phial," Walsh wrote to Franklin, "will hear with pleasure that its laws prevail in animate phials."[35]

Another thing that Walsh demonstrated, although it was not actively pursued at the time, was that the electric eel could sense electricity as well as generate it. When one or more conductors were placed in the water of an eel's tub, the fish ignored them. But if they were connected in a circuit outside the water, "the animal becomes agitated, and rushes to them, and brings the extremity of its head to one end of this conductive arc as if he would like to smell it, he provokes the electric discharge." But never mind an "electric sense": in the mid-1770s, many people were skeptical of the idea that fish could produce electricity at all. One of these holdouts was instrument maker William Henly, who specialized in devising electrometers. Henly tried building an artificial model of the torpedo out of conductive materials, but it proved incapable of attracting or repelling chaff or producing a perceptible sensation of any kind. (It is not known whether Henly was familiar with Robert Turner's earlier attempt to make an artificial torpedo using an electrified flounder.) Henly's explanation for the failure of his device was based on an argument originally made by Albrecht von Haller. Electricity was known to flow through conductors in such a way as to eliminate any imbalance, and living tissue was known to be conductive, so how could any stable electrical imbalance build up in living tissue? Worse, since the fish's habitat was also conductive, by this reasoning it should also be impossible for there to be an imbalance between fish and environment. Henly remained unconvinced of the possibility of fish electricity until he saw Walsh's 1775 demonstration with the electric eel *Gymnotus* in London.[36]

Following Henly's attempt, Henry Cavendish also tried to create an artificial torpedo, with more success. Cavendish noted that Walsh's experiments seemed to show that the creature was able to throw "at pleasure a great quantity of electrical fluid" from organs on its upper surface to those on its lower, or vice versa. Furthermore, people who touched the fish underwater received a shock, even though the water itself would seem to provide a shorter circuit than the person's body. Cavendish argued that rather than taking the shortest path, electricity passed through any number of circuits simultaneously, with more electricity flowing through those circuits that had less resistance. If a person held a thick wire with both hands while electricity was sent through it from a Leyden jar, his body would form a parallel circuit, but unless the jar was quite large and highly charged, he wouldn't feel the shock. Since a thick wire has little resistance, by Cavendish's logic the majority of the electricity would flow through it and only a tiny proportion through the person's body. Cavendish's experiments suggested that iron conducted electricity about 400 million times better than rain or distilled water, and that seawater was about one hundred times more conductive than rainwater. The difference in conductivity between air and water could explain Walsh's observation that the shock of the torpedo was perceived to be about four times stronger in air than in water: the person's body provided the path of least resistance when the fish was held out of the water.[37]

Rather than thinking of the body of a strongly electric fish as a simple conductor, as Henly had done, Cavendish created a more sophisticated conceptual model. He imagined that the fish's body acted as a series of Leyden jars connected in a battery. Each of the jars would be weakly charged, but together they could store a lot of electricity. This idea was supported by the work of anatomist John Hunter, who had been given torpedoes to study by John Walsh. In 1773, Hunter had shown that the electric organs of the torpedo were composed of hundreds of columns, each further subdivided by horizontal membranes to create tiny fluid-filled cells. At least superficially, these cells resembled miniscule Leyden jars. Cavendish's Leyden jar model also explained how the torpedo could shock without producing a perceptible spark, because a large battery of weakly charged Leyden jars wired together could still give a powerful shock, even though it could not create a visible or audible spark. The larger such a battery was, the smaller the gap that its spark would jump. If the tiny pockets of fluid that Hunter had identified in the torpedo's body were actually operating as weakly charged Leyden jars, there were many of them indeed.[38]

Cavendish shaped a piece of wood to the approximate dimensions of a torpedo, albeit one with a 40-inch-long handle protruding from its snout. He cut a groove into the handle to hold a long glass tube, fed wire down the tube, and soldered it to a pewter plate on the body of the model fish. The fish part was then covered with a piece of sheepskin leather and thoroughly soaked in seawater. He then placed his hands near the model fish, had his assistant wire the model to a battery of forty-nine Leyden jars, and received a shock. When suspended in air, the shock from the artificial torpedo felt just like a real torpedo; when placed in the water, however, the apparatus gave off a barely perceptible shock. Cavendish concluded that the wood he was using to model the fish's body was not conductive enough, and he improved his model by replacing that part with several thick pieces of leather that had been cut into the shape of a torpedo and fastened together in layers. The new model shocked in a way that felt much closer to the actual torpedo.[39]

Like other eighteenth-century natural philosophers, Cavendish used his own body as a component in the electrical circuits he was exploring. Working with the artificial torpedo, he noted that there was "a considerable difference between the feel of [the shock] under water and in air. In air it is felt chiefly in the elbows; whereas, under water, it is felt chiefly in the hands, and the sensation is sharper and more disagreeable." Cavendish differed from other "electricians" of the time, however, in the degree to which he was able to calibrate his own perceptions. He did this by using devices with variable resistance. These were thin glass tubes with a wire electrode inserted in either end, filled with a conducting solution. The resistance of the device could be reduced by moving the electrodes toward one another in the tube and increased by moving them apart. When different solutions differed in conductivity, the wire electrodes would have to be pushed closer together in one device than in the other in order for each to offer similar resistance. Cavendish could then compare the conductivity of two different solutions as follows. He would put each solution into one tube, charge several Leyden jars to the same degree, then discharge them through first one tube and then the other. Feeling the shock each time, he could adjust the wire electrodes until the sensations seemed the same. The conductivities he was able to measure in this manner were not only consistent with one another, they were also very close to measurements made decades later by other observers.[40]

No matter how fine or well-calibrated an instrument, however, Cavendish's perceptions alone were not sufficiently convincing for the context in which he and his colleagues produced scientific knowledge. The historian Peter Dear argues that the scientific societies founded in the mid-1600s, including the Royal Society to which Cavendish belonged, embraced an attitude toward natural knowledge that was grounded in the observation of specific events. Reports to scientific societies took the

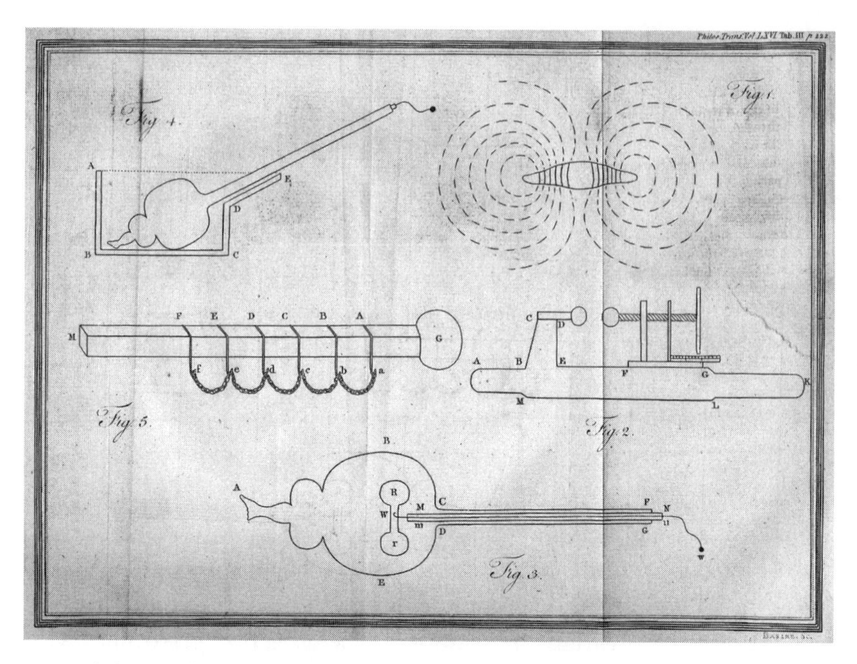


Cavendish's artificial torpedo discharged electric shocks that were qualitatively the same as those from the Leyden jar. From Henry Cavendish, "An Account of Some Attempts to Imitate the Effects of the Torpedo by Electricity," *Philosophical Transactions of the Royal Society of London* 66 (1776): 196–225.

form of statements about how something had happened, or was experienced, on a particular occasion. This was in marked contrast to earlier, scholastic practice, in which statements tended to be generalizations about the typical behavior of something. Written accounts of experiments tended to conform to a pattern: first an impersonal description of the experimental setup and the procedure to be followed, then a detailed description of what happened on the occasion that the experiment was actually performed. The identity and social status of the experimenter helped to determine how the report was received. Gentlemen were considered more trustworthy observers, but their putative inferiors were thought to be less likely to be biased toward any particular hypothesis. Laboratory objects were the most credible of all; in the words of Robert Boyle, "inanimate bodies . . . are not capable of prepossessions, or giving us partial informations."[41]

One way to increase the authority of a report was to perform the experiment in the presence of witnesses. Consequently, Cavendish brought several men into his laboratory on Saturday May 27, 1775, to experience his artificial torpedo. These included Joseph Priestley, the anatomist John Hunter, Timothy Lane, and Edward Nairne, who had each built some of the equipment that Cavendish was using, and Thomas Ronayne, who had corresponded with Benjamin Franklin on the subject of atmospheric electricity but was skeptical about the possibility of electric fish. For if such creatures existed, Ronayne didn't see anything preventing the occurrence of underwater lightning storms. "Indeed," he said, ". . . when a Gentleman can so far give up his reason as to believe the possibility of an accumulation of electricity among conductors sufficient to produce the effects ascribed to the Torpedo, he need not hesitate a moment to embrace as truths the grossest contradictions that can be laid before him." But in Cavendish's laboratory all felt a shock, including Ronayne. Cavendish concluded that the shock of the torpedo was compatible with what was known about electricity.[42]

Cavendish's biographers write that his approach to the torpedo resulted in a series of fundamental questions about the nature of electricity, its manifestations, and the possibilities of storing, transmitting, manipulating, and measuring it. The fact that the torpedo seemed incapable of creating a visible or audible spark or of attracting chaff suggested to many natural philosophers that the fish's mechanism could not be electrical in nature. In Cavendish's hands, however, the evidence took on new significance. He argued that one must distinguish between the intensity of electricity and the quantity. Using sophisticated electrometry, he was able to show that a battery of Leyden jars would behave differently from a single jar that had an identical charge. The shock was calibrated to feel the same in both cases, but only the latter created a sensible spark. This reasoning could then be transferred from the artificial torpedo to the real one. As the historian Simon Schaffer notes, by controlling the behavior of a model, experimenters felt justified in making bolder claims about natural phenomena. Eighteenth-century natural philosophers also created "thunder houses," tiny models of churches and houses loaded with gunpowder so that they would burst into flames when struck with the "lightning" from a Leyden jar. Animal bodies played a strange part, and a much less direct one, in some of these demonstrations, as charged clouds were occasionally represented by a metal-plated placenta or "copper-coated bullock's bladder." Animal electricity, atmospheric electricity, and model-scale demonstrations were all seen to be manifestations of the same physical phenomenon.[43]

One consequence of seeing the world in terms of apparatus is that people are able to co-opt properties or abilities of other animals and put them to use in new contexts. We have several strategies for doing this. We can try copying the form of something, extracting the material of interest, or copying its function. In the case of animal electricity, neither form nor substance were of much use, but the Leyden jar served as a functional analog. This new electrical technology made it clear that animal electricity was the same kind of thing as atmospheric electricity. It also allowed experimenters to begin to quantify their own sensations, and those of other people, in a way that provided a deeper understanding of how electricity worked. Not coincidentally, this new mastery of electricity would provide a deeper understanding of how people worked, too.

THREE

# Electrophysiology

## Nerves

As physicians and natural philosophers explored the physiological implications and therapeutic applications of electricity, they came to new understandings of sickness and health, life, and death. While working with preparations made from the flesh of living or recently killed animals, they also discovered that it was possible to build "artificial electric organs." These devices were convenient because they delivered a steady electric current, unlike the Leyden jar, which discharged its contents in a single blast. Artificial electric organs are now ubiquitous, having been completely dissociated from their origins in animal tissues: this is the bloody history of the battery. We can also see it as part of the natural history of humankind, a story of animals whose particular genius lies in disassembling and refashioning everything around them, living or dead.

Whatever the mechanism responsible for nervous and muscular action, by the mid-seventeenth century it was becoming clear that it could not be due to the pneumatic flow of animal spirits posited by the Galenic model. In the early 1660s, the Dutch naturalist Jan Swammerdam began a series of dissections of and experiments with frogs. He confirmed the observation that cutting or touching a nerve with a knife caused the associated muscle to contract, and he showed that contraction of the limbs also followed when the animal's spinal marrow was irritated. He rejected the idea that nerve fibers were hollow and carried animal spirits and devised a clever apparatus to demonstrate this. Swammerdam started by dissecting a large thigh muscle from a frog without removing the adherent nerve. He then placed the muscle and nerve in a sealed glass vessel with a silver wire running from the nerve through a small copper loop to the outside of the container. By pulling on the wire, he could mechanically stimulate the nerve. The glass vessel had a thin tube projecting from one end. A drop of water placed near the end of the tube served as an indicator. If the volume of the muscle changed when the nerve was stimulated, it would displace air in the vessel and the thin glass tube, thus moving the drop of water in one direction or the other. When Swammerdam stimulated the nerve, the muscle contracted, visibly swelling and changing shape but not volume. He thus concluded that the muscle was not being inflated by animal spirits. In fact, the water droplet was drawn inward slightly when the muscle contracted. He drew the further conclusion that the motion of bodies was due to perpetual irritation of the nerves. Swammerdam demonstrated his frog neuromuscular preparation widely, but his work was not published until the 1730s.[1]

The concept of irritability was elaborated the English physician Francis Glisson and by Albrecht von Haller. Glisson believed irritability to be a property of all tissues, not only of nerve and muscle. He argued that this universal irritability took three forms. Motion could be initiated by the soul, arise as a result of an external stimulus, or be intrinsic to the body and beneath the threshold of consciousness. Following in Glisson's footsteps, Haller did extensive, gruesome experimentation. He started with living dogs, cats, goats, rats, rabbits, and other animals of various ages and tried "touching, cutting, burning, or lacerating" various parts of them to determine which were sensible, using the cries of the animal as one would the gauge of an instrument. He concluded that fat and tendons were not, that skin was, and that the sensitivity of muscular flesh depended on the nerves. When isolated from nerves, muscles would contract in response to physical stimulation, although the animal was apparently not sensible of this. Haller thus distinguished sensibility, a property that he attributed to the nerves, from irritability, the contractile response of muscular fiber to physical stimulation. As a surgeon, Haller had an opportunity to confirm some of his findings while operating on people, but most of his experience came from the vivisection of 190 animals. Haller's descriptions of these experiments, published in 1753, are revolting. He justified them as "a species of cruelty for which I felt such reluctance, as could only be overcome by the desire of contributing to the benefit of mankind." In his own defense, Haller also noted that "persons of the most humane temper . . . eat every day the flesh of harmless animals without any scruple."[2]

The network of nerves was clearly capable of "feeling," but what mechanism allowed this? In the mid-seventeenth century, some scholars had suggested that the nerves vibrated like the strings of a musical instrument, an idea developed by Isaac Newton early the following century. In his work on optics, Newton had observed that pressing a fingertip against his closed eye resulted in the perception of a ring of colors, like a peacock's tail feather. If he held the finger in place, the colors vanished. He compared the effect to that of throwing a stone into still water. If a burning ember were swung in a circle, one saw a circle of light rather than a moving point. Newton argued that wavelike motions were being excited in the eye by light and died down gradually in its absence. He added a "General Scholium" to the second edition of his *Principia* (1713) that described a "subtle spirit" that pervaded all bodies, allowing them to attract or repel one another at a distance. "And all sensation is excited," he wrote, "and the members of animal bodies move at the command of the will, namely, by the vibrations of this spirit, mutually propagated along the solid filaments of the nerves, from the outward organs of sense to the brain, and from the brain into the muscles." Andrew Motte's English translation of Newton's work (1729) described this spirit as "electric and elastic," a phrase that does not appear in the Latin original, leading to a subsequent debate about the degree to which Newton intended this reading. From his unpublished drafts and marginalia, it is clear that Newton thought of this force as a "spiritus electricus," which is how he was read by his contemporaries.[3]

Haller rejected the possibility that the nerves might be mechanically vibrating. Nerves were not under tension at either end, and in many cases they were firmly attached to other tissues such as arteries. But the Leyden jar was coming to prominence at the same time as Haller was doing his experiments on irritability, and it was clear that electrical shocks resulted in strong muscular contraction. Given Newton's endorsement of an electric spirit pervading all things, one possibility was that nervous fluid was a kind of electricity. Haller rejected this idea. From first-hand experience, he knew that tissues were electrically conductive and that nerves were not insulated. He did not see how it could be possible for electrical fluid to remain confined in the nerves. (Haller also used a version of this argument against the possibility of fish electricity: no imbalance of electrical charge should be able to build up in the fish's body or between its body and environment.) Furthermore, "a ligature on the nerve takes away sense and motion, but cannot stop the motion of a torrent of electrical matter." Haller's conclusion was that nervous fluid, while subtle, must be "more substantial than heat, aether, electricity or magnetism."[4]

As with electricity, the parts played by brain and soul in nervous action were uncertain, too. One suggestive line of evidence came from the involuntary motions made by decapitated animals or those with broken necks. (Violent animal deaths were a familiar experience for most people before industrialized agriculture sequestered animal slaughter.) Glisson had suggested that these twitches were similar to a class of movements that were not willed but were due instead to some lower form of perception. Robert Whytt showed that the spinal cord was crucial to these movements because nerves communicated only at their termination in the brain or spinal cord. Since the spinal marrow was necessary for involuntary motion, the movements of decapitated animals could not be due to tissue irritability as Haller proposed. Whytt attributed it to a "sentient principle" instead. Considering the writhing of a decapitated snake, Whytt concluded that the carcass in some sense still lived, must still feel, still be "animated by a sentient principle."[5]

Haller disagreed. Irritability could not depend on the soul because "this quality still subsists after the seat of the soul is removed, or its commerce with the body quite intercepted." But Haller was a pious man. Even if a soul was not required to explain the phenomenon of irritability, he certainly did not want to dispense with souls altogether. Much to his dismay, that was exactly the conclusion drawn from his work by Julien Offray de La Mettrie. In his infamous *L'homme machine*, La Mettrie pushed Descartes's claim that animals are machines to its logical conclusion: since people are animals, people must be machines, too: "To be a machine, to feel, think, know good from evil like blue from yellow, in a word, to be born with intelligence and a sure instinct for morality, and yet to be only an animal, are things no more contradictory than to be an ape or parrot and know how to find sexual pleasure . . . Thought is so far from being incompatible with organized matter that it seems to me to be just another of its properties, such as electricity, the motive faculty, impenetrability, extension, etc." Haller's irritable fibers, each capable of independent motion, served as "the springs of the human machine" in La Mettrie's argument. The pupils enlarge and contract in dark and light, pores close in the cold; the stomach heaves when poisoned; the heart beats while asleep; the lungs work like bellows; the sphincters of bladder and rectum open and close without conscious intervention. To La Mettrie, even "feelings, pleasures, passions, and thoughts" worked the same way, "for the brain has its muscles for thinking."[6]

The question of the relation of mind to brain remains vexed to this day, needless to say. In the mid-eighteenth century, the physician David Hartley attempted to ground psychological associations in a theory of nervous action that drew on Isaac Newton's account of neural transmission as vibration. Hartley was aware of the arguments of Haller and others against the possibility that whole nerves were vibrating like musical strings, an idea he, too, found absurd. Instead he espoused a combination of a subtle medium composed of vibrating particles and microscopic vibrations of the particles that made up the nerve itself. These oscillations had to be small enough relative to the scale of the body and its tissues that they would not perturb it mechanically. One analogy he proposed was the transmission of an electrical charge through a hempen string. In Hartley's theory, vibrations in the aether were transduced (to use a twentieth-century verb) when they excited synchronous vibrations in the nerve. These neural vibrations could indicate the degree, kind, location, and direction of an external event. For Hartley, association between ideas was also due to synchrony between vibrations. If the research practices of natural philosophers like Henry Cavendish suggested that sensation and perception were instruments that could be quantified and calibrated, then the same mathematics that was used to analyze a clockwork universe might be extended to the world within.[7]

## Moral Theater

Neither vivisection nor electrical fire was confined to laboratories. Both practices made their way through more public circuits, including schoolrooms, salons, and the popular press. The historian Anita Guerrini argues that public demonstrations of animal experiments "forced [their] audience to consider the meaning and purpose of life and death. Since animals acted as stand-ins for humans, the question of their moral status and their experience of pain were implicit in the act of experimentation." Witnessing the suffering of innocent animals was thought by some philosophers to lead to compassion. Animals played a role in what Guerrini calls "moral theatre," "a demonstration touching on the interconnections between religion, ritual, and secular society." For example, the anatomy theater in Leiden was located in a former church, decorated with moral images and mottos as well as preserved animal and human remains. Since postmortem dissection was one fate for executed criminals, some of the skeletons were labeled with the crimes their former owners had committed. Audience members could not help but be confronted with thoughts of their own mortality while watching a dissection take place there. In Edinburgh, Alexander Monro *Primus* made extensive use of live animals in his lectures to medical students and auditors, flaying, intubating, inflating, injecting, and suffocating the poor creatures to make various points. Although he recognized that the practice was cruel, he believed that it was necessary, and justified by the ends to which the resulting knowledge could be put: the advancement of natural theology, natural philosophy, forensics, art, and medicine.[8]

Another form of animal experimentation that contained a fair degree of theater and one-upmanship (and one that would eventually have grisly consequences for human beings, too) was the practice of electrocuting larger and larger animals with batteries of Leyden jars. In France, Abbé Nollet took pairs of cats, pigeons, chaffinches, and sparrows, weighed them and separated them into individual wooden cages. He chose one of each pair to electrify for five or six hours, then weighed them again, finding that the electrified animal was lighter than its matching control. He then repeated the experiment, switching subject and control, with the same results. The electrified animal was consistently lighter than the other, an outcome that Nollet attributed to an increase in insensible perspiration. (Young men and women subject to the same regime reported no discomfort, but they were more tired and hungry at the conclusion of being electrified. The abbé feared that their clothes were absorbing sweat and spoiling his measurements.) Nollet moved on to more lethal pursuits. He killed a sparrow with electricity and then dissected it, noting that its blood vessels had burst and that its injuries resembled those caused by lightning. In Danzig, Daniel Gralath was one of the first people to successfully replicate Kleist's experiments with the electric phial. Gralath started by electrocuting beetles and worms, then moved to small birds, eventually reaching the limits of his ability to take life with electricity when he failed to kill a goose. Benjamin Franklin took it further, electrocuting guinea fowls and a turkey, and Joseph Priestley dispatched a range of animals from shrew to dog. That was about the most that could be accomplished with Leyden jars. The electrocution of larger animals—including humans—would have to await nineteenth-century technology.[9]

Practices such as vivisection and animal electrocution seem horrible to us, but they need to be understood in context. Although some of the natural philosophers who performed animal experiments were Cartesians, they believed that animals were capable of feeling pain, whether they were "machines" or not. The experimenters typically justified causing animals to suffer by claiming that it would ultimately reduce human suffering. At the same time, the torment of animals was also on display in much less edifying "entertainments" like bull-baiting and dog- and cockfighting. For that matter, it was also possible to watch men beat each other with cudgels until one combatant drew blood from the other's head, and executions would be held in public in England until 1868. It goes without saying that people can still be seen beating, torturing, and killing one another and tormenting and killing animals on television or YouTube, in the guise of news, sport, or fiction. By the mid-seventeenth century, however, the tide of public sentiment was in the process of turning against cruelty toward animals, including the subjects of experimentation, and more humane practices, like keeping pets, were on the rise.[10]

Displays of electricity, while more-or-less painful, were also fashionable and frequently titillating. Sparks were made to jump from lips, fingers, hair, and petticoat hoops in demonstrations with names like "beatifying electricity" and "Venus Electrificata, the electric kiss." Leipzig professor Johann Heinrich Winkler made a name for himself by electrifying a servant then offering the man a glass of brandy. Sparks from the servant's tongue ignited the brandy to the amusement of everyone present, probably excepting the servant.

In order to replicate effects like these, participants needed to learn how to hold unfamiliar postures and perform new physical routines. Simon Schaffer argues that the difficulty of replicating particular electrical demonstrations confronted public lecturers and showmen with a problem that sociologist of science H. M. Collins calls "the experimenter's regress." The outcome of an experiment is supposed to tell you something about the way that the world works. That is well and good if you know what the outcome should be. But suppose the phenomena that you are investigating are completely novel; then it is not clear whether a particular experiment should result in the detection or the nondetection of the phenomena in order to be considered a success. Electric fire was novel. In order to determine if a demonstration was properly set up, the showmen put their own bodies into the circuits. "A mid-eighteenth-century electrician knew that a body was electrified if he could draw a spark via his finger," Schaffer writes. "He followed the fire's path by tracing the shock through his body." Like Cavendish, these public performers calibrated their own perceptions of electric shock to increase the reliability of their instruments and practices.[11]

The spread of electrical fire through more popular venues was due in part to a series of articles published in the *Gentleman's Magazine* beginning in the mid 1740s. The magazine had thousands of paid subscribers in Britain and the colonies and was increasingly devoting space to medical and scientific news of a practical bent. The April 1745 number contained an anonymously authored five-page "historical account of the wonderful discoveries, made in Germany, &c. concerning Electricity." The historian John L. Heilbron has argued convincingly that thanks to this article it is likely that Benjamin Franklin first learned how to work with electricity from Albrecht Haller. Heilbron shows that the article in the *Gentleman's Magazine* was derived from a French review in the *Bibliothèque raisonnée*, which was, in turn, derived from German reviews written by Haller in the *Göttingische Zeitungen*. Haller may also have been the author of the French language review. Given that it took six months to ship, the April 1745 edition of the *Gentleman's Magazine* would have arrived at

Philadelphia in the late fall, when Franklin and his group would have found electrical experimentation to be a welcome indoor activity. Haller's descriptions of experimental setups were clear enough to enable novices to obtain good results, and they shaped Franklin's subsequent demonstrations of plus and minus electricity and his exploration of lightning. The Philadelphians also improved on the electrical kiss and other games that Haller depicted in glowing terms.[12]

In 1746, readers of *Gentleman's Magazine* learned of Musschenbroek's letter to Réaumur describing his dangerous experience with the Leyden jar: "the commotion he felt was like a clap of thunder." Another piece described experiments using electrically charged glass tubes to fire gunpowder and ignite flammable spirits in a spoon. Those who wished to replicate the results that they read about often did so and reported back to the magazine with further details. Toward the end of the decade, articles about new medical uses for electricity began to appear in *Gentleman's Magazine*, and within a few years readers could also learn of various encounters with strongly electric fish. An anonymous account of a voyage to Chile included a description of a torpedo caught off the Mexican coast, comparing it with the eels observed in Guiana by Réaumur (who actually didn't, but who corresponded with people who did). A couple of years later, a biographical sketch of Réaumur mentioned his work with torpedoes and his mechanical theory of their action. In 1769, a review of Edward Bancroft's *Natural History* described Bancroft's experiments with electric eels and argued that these fishes' ability to shock was due to electricity rather than vibration. In the early and mid-1770s, the magazine covered Walsh's experiments. Stanley Finger and Ian Ferguson suggest that the editors of the *Gentleman's Magazine* were most interested in making the news lively and readable and didn't pay much attention to details they thought would confuse or bore their readers. Cavendish didn't make the cut; those who were interested were directed to the *Philosophical Transactions* to learn about his attempts to imitate the torpedo's effects with electricity.[13]

Salon participants who played games with electrical fire temporarily developed small red spots on their hands, something that came to the attention of physician and professor Johann Gottlob Krüger in 1744. Since electricity also elicited involuntary movements, Krüger hypothesized that it might be used to give victims of palsy the ability to feel and move again.

He and his student Christian Gottlieb Kratzenstein began experimenting. Kratzenstein found that mild electrification restored movement to the fingers of two of his patients. Anecdotal reports from people who had played with electricity suggested that heart rate might increase as a result of electrification, something Kratzenstein was also able to confirm. Since being electrified seemed to make people tired, Kratzenstein thought it might be a helpful remedy for insomnia. News of his electrical treatments soon spread through Europe. The new availability of inexpensive and portable electrical machines led to more medical experimentation—often reported in the pages of *Gentleman's Magazine*—leading in turn to improvements in technology. Wittenberg professor Georg Matthias Bose, inventor of many popular electric pastimes, showed that water would spray from an electrified siphon but would only drip from one that was not electrified. Quick to spot a mechanical analogy, Bose also showed (as Abbé Nollet later reported) that blood "stream[ed] off more quickly" from the opened vein of an electrified patient and that the "blood drops appear[ed] luminous like fire."[14]

## Electrotherapy Redux

For a while it seemed as if practically any ailment might be treated with electricity. In Vicenza, the marquis Luigi Sale tried zapping his sleepwalking servant with an electrical machine to wake him up. The man was something of a medical legend, already the subject of several booklets that described some rather extreme but ultimately unsuccessful attempts to awaken him during one of his somnambulant episodes. He was apparently so grateful for Sale's cure that he tearfully blessed the machine that brought him some relief. In Venice, Gianfrancesco Pivati sealed medicinal substances like camphor, opium, and Peruvian bark in glass cylinders and then electrified these "medicated tubes." Patients who inhaled the air in the vicinity of the tubes were, according to Pivati, healed by the medicine transmitted through the pores of the glass by electrical fluid. Although the electrified tubes did not smell of anything themselves, patients who inhaled near them or handled them were said to absorb the medicines in such quantity that their sweat, bedding, clothes, and even combs were "perfumed" by it. Pivati's results were confirmed by one physician in

Bologna, only to be discredited by another in Venice the following year. News of the medicated tubes was sent to the Royal Society in 1748 and appeared in the *Gentleman's Magazine* the following year.[15]

In Geneva, Jean Jallabert reported successful treatment of paralytics with electricity. He first used a Leyden jar to stimulate involuntary contractions in the muscles of his own arm, learning where to apply the shocks for maximum benefit. Jallabert then began treatment of a man named Nogues in public, drawing sparks from his paralyzed arm for two hours each morning and two to three each afternoon. After a month and a half of this regimen, Nogues retained the use of his arm, and Jallabert spread the news throughout Europe. In Paris, Jallabert's friend and correspondent, Abbé Nollet, was electrifying paralytics at the Hôpital des Invalides in the late 1740s. Nollet was unable to effect a cure, however, and concluded that electrification could only lead to involuntary movements. He was also unable to get satisfactory results from the use of medicated tubes. The abbé decided to make his way across the Alps into Italy in the summer of 1749 to witness the treatments firsthand. Alas, none of these miraculous cures seemed to work while he was present, and he widely denounced the "charlatans" promulgating them and bemoaned the suggestibility and superstition of the sick, especially when the sufferers were children, servants, or members of the lower class. Simon Schaffer argues that calling lay audiences suggestible or superstitious served natural philosophers in two ways: not only did it make natural philosophy more appealing to hoi polloi, but it "could also be used to explain away its failures and pathologies."[16]

Possible applications of electrotherapy were enthusiastically investigated in the 1750s. The first English language textbook on the subject was published in the middle of the decade. It described three kinds of treatment: electrification, drawing sparks, and shocking. The first involved insulating the patient and connecting him or her to an electrical machine to replenish electricity that had supposedly been lost. In the second case, an excess of electricity was drawn from an insulated patient by means of a metal rod. In the third treatment, a Leyden jar was discharged through some portion of the patient's body. In each of the treatments, the human body was conceptualized as a kind of a container for electrical fluid. Therapeutic applications required careful administration, which stimulated medical instrument makers to find ways to measure and regulate the amount of electrical fluid that would be discharged in a given shock. Results piled up within the span of a few years: physicians showed that muscular tissue contracted when electrified; they suggested that it might be used to cure hemiplegia; they claimed that it did cure epilepsy, paralysis, spasms, and St. Guy's dance; they used it to suppress menstruation; they reported success treating rheumatism, toothache, hypochondria, neuralgia, intermittent fever, lockjaw, and paralysis of the optic nerve. Dozens of medical men reported successful treatment of paralysis. Toward the end of the decade, the Reverend John Wesley, founder of Methodism, was moved to write a treatise subtitled *Electricity Made Plain and Useful, by a Lover of Mankind and of Common Sense*. In it, he listed more than one hundred cases of electrical cures of everything ranging from agues and St. Anthony's Fire to surfeit and wen.[17]

In the 1750s and 1760s, medical experimentation with electricity was also pursued in the colonies, where easy access to strongly electric fish allowed electrical machines and electrical animals to be used interchangeably for therapy. Reports from Barbados suggested that people with gouty limbs obtained relief within minutes when they touched electric eels. Although using strongly electric fish therapeutically was a traditional aboriginal practice, these patients were presumably not slaves or Indians. Gout was a disease that followed from the steady consumption of meat and fortified wine, in abundance or excess. In the Dutch colony of Essequibo, Frans Van der Lott (who was either a surgeon or a jurist: the sources disagree on this point) used electric eels to treat chickens that were suffering from cramped claws and could not walk or forage to survive. They screeched dreadfully when he touched their legs to an eel swimming in a tub of water but recovered their mobility. Van der Lott then turned his attention to an Indian man who was paralyzed from the waist down. He had two assistants hold the man up by his arms and applied the fish against his knees. The shock almost knocked all three of them to the ground. After three repetitions, the Indian was apparently able to walk without crutches back to the plantation from which he had been carried.[18]

Slaves proved to be an especially popular choice of subjects for would-be colonial electrotherapists, no doubt because their consent was not obtained before experiments were conducted on them. Van der Lott reported on several patients that Abraham van Doorn, ex-counselor of

Essequibo, attempted to treat. One, a black slave boy about eight or nine years old, had crooked arms and legs. Every day, Van Doorn threw the boy into a tub containing a large electric eel, "which shocked the boy so powerfully that he crept out on all fours" or, failing that, was pulled from the water by an assistant who was shocked in the process. The boy supposedly recovered, although his shinbones remained malformed. Van Doorn also threw a feverish slave boy into a tub with an electric eel to cure him, something that had apparently helped an Indian boy of his acquaintance. Slaves who complained of bad headaches were instructed to place one hand on their head and the other on an electric eel. Demonstrations that involved shocking a circuit of people holding hands were performed with slaves in colonial settings. In all cases, the results were wonderful, at least from the perspective of the Dutchmen who reported them. The physician Godfried Willem Schilling, who lived in Surinam, also used slaves in his experiments with electric eels. In 1770, he reported that electric eels were attracted to lodestone, which adhered to them and took away their power to shock. Several prominent natural philosophers subsequently failed to replicate Schilling's findings.[19]

The historian James Delbourgo notes that colonial accounts of strongly electric fish often compared the bodily sensibility of different races, either explicitly or implicitly, suggesting equality in some cases and hierarchy in others. In the Persian Gulf, the seventeenth-century traveler Engelbart Kaempfer wrote of "A certain African . . . [who] boldly lifted up the torpedo repeatedly and held it without any sensation of horror," apparently because he held his breath while doing so. On the River Plate, the Jesuit Ramón M. Termeyer experimented extensively with electric eels. He tried holding his breath while touching the animal but was shocked anyway. He used the fish to shock other fish, chickens, a cat, and a chain of four dogs, comparing the effects to those of the electrical machine and Leyden jar. Termeyer reported that there was a Mocoví Indian boy who could touch the electric eel without being shocked; he did feel the shock of Termeyer's Leyden jar, however. Another Jesuit missionary serving in what is now Bolivia and a correspondent of Termeyer, found hundreds of people that were susceptible to the shock of the electric eel, save one. Henry Collins Flagg, a South Carolinian who was stationed at Essequibo in 1782, was informed that "some Indians and negroes" could handle the fish without being shocked. He also knew a lady who was capable of the same feat, something he attributed to her constitution. "There was more than an air of the prodigious to this account," Delbourgo writes, "were ladies not the most sensible of creatures rather than the least?"[20]

While introducing electricity to their readers in 1745, the *Gentleman's Magazine* celebrated the gendered roles of electrical demonstrations like the electrifying Venus: "The ladies were sensible of this new privilege of kindling fires without any poetical figure, or hyperbole, and resorted from all parts to the public lectures of natural philosophy, which by that means became brilliant assemblies." Bose, inventor of many of these entertainments, liked to distinguish the powerful "male" electricity that sparked and crackled between metals and the bodies of animals from the pale "female" luminescence that accumulated around charged objects during a demonstration like the "beatification."

The friction of rough hands on smooth spinning globes inspired pornographic satire. In 1780, George Graham's Temple of Health and Hymen in London offered couples a medico-electric and magnetic "Celestial Bed" to promote fertility. Around that time, the word "spark" came to be used for both for an electrical discharge and for a young lover. And the electric eel—most powerful and phallic of the strongly electric fish—was put on display in London in the late 1770s. The actual fish died within a few months, but the idea of an eel "which made the very dullest rise" became a mainstay of Georgian bawdy and erotic fiction. Historian Karen Harvey notes that experiments with electricity were the subject of a number of erotic texts published in the 1770s. One example, *The Torpedo* (1777) was an erotic poem that likened the female body to the torpedo, the male to the eel, and used electricity as a metaphor to comment on "desire, potency and fertility." Male and female were sexually equivalent, "but the female torpedo emerges as a much more powerful force with 'sparkling fires' and 'stronger flames.' Indeed, not only is the electricity of the torpedo more powerful than that of the eel, but the eel's electricity 'From the TORPEDO came.'"[21]

## Galvani and Volta

By the last quarter of the eighteenth century, experimenters were routinely vivisecting animals in the laboratory and in public to demonstrate the electrical nature of "nervous fluid." Earlier studies (such as those on irritability and sensibility) had relied on doing brute damage to living creatures to see what happened. Now the apparatus was becoming more intricate and more mechanized, although the quality of suffering probably did not change much for the animal subjects. The newer procedures and preparations treated the body parts of living or freshly killed animals as interchangeable components, combining them with one another and a variety of other inanimate parts and instruments. More use was made, too, of multiple copies of a given animal, limb, or tissue.

In 1780, Luigi Galvani began a series of experiments on frogs to determine how "motion, sensation, blood circulation, even life itself" depended on electric fluid. Galvani had studied medicine and surgery at the University of Bologna, where he had also learned about Benjamin Franklin's research on lightning and had become proficient in the use of electrical instruments. Galvani was interested in the therapeutic use of electricity and published a paper (now lost) on Albrecht von Haller's theory of irritability, which held that animal spirits didn't cause muscular contractions but served as a stimulus for them. Galvani was instead a proponent of the opposing theory that muscular contractions were due to electrical fluid. Within a few months of beginning work with the frogs, Galvani arrived at a demonstration that would later be known as his first experiment. Cutting one of the animals in half, he placed a metal wire on the spinal canal of the lower portion. The creature's limbs then contracted violently if the crural nerve was touched with a lancet while a nearby electrostatic machine or Leyden jar emitted sparks.[22]

To investigate natural atmospheric electricity, Galvani set up a frog preparation on a stormy evening, with a long metal wire inserted into the spine and extending toward the sky. Sure enough, thunder and lightning were accompanied by strong muscular contractions. Aided by his wife, Lucia Galeazzi, and his servants, Galvani inserted brass hooks in the spinal cords of his dissected frogs and hung them from the iron fence surrounding the family's sunken garden. There he observed that the leg muscles contracted whether there was an electrical storm or not. This led him to wonder if the contractions were really due to changes in atmospheric electricity, but he was unable to systematically relate the two phenomena. Taking his frog preparation indoors and placing it on a metal plate, Galvani found that the muscles contracted when he touched the end of the brass hook to the plate. He replicated the new experiment many times, eventually satisfying himself that the strength of the contraction depended on which metal was used. When he substituted insulators or weak conductors for metal the muscles didn't contract. This surprising result led Galvani to consider the possibility of an inherent animal electricity, and his hunch was strengthened when he realized that "when the phenomenon of contraction occurred the flow of very tenuous nervous fluid from nerves to muscles resembled the electrical flow that is discharged in a Leyden jar." Galvani tried putting himself in the circuit, too, dangling the frog preparation by its hook from one hand so that the feet brushed a silver box. When he touched the box with his other hand, the leg muscles contracted.[23]

Galvani's work evoked a strong response when he published it in 1792. He suggested that animal bodies contain an innate electricity, comparing individual muscle fibers to Leyden jars, nerves to conductors, and the muscle as a whole to a battery of Leyden jars. The muscles could thus hold animal electricity in "disequilibrium" until internal or external conductors allowed the electric fluid to flow through a circuit, resulting in contraction. News of Galvani's research spread quickly, leading other natural philosophers to replicate his experiments.

One who did so was Alessandro Volta, professor of physics at the University of Pavia. Volta had been aware of the idea of animal electricity for at least a decade. In a letter of 1782 he wrote, "In order that we may be entitled to speak of animal electricity, we must find a kind of electricity essentially linked with life itself, and inherent in some animal function. Now, does such electricity really exist? Yes, it has been discovered in the torpedo and in the electric eel of Surinam." Volta initially believed that Galvani was correct to attribute the contractions to intrinsic animal electricity, but he came to doubt the explanation as he experimented further. When he created a length of wire made from two different metals and pressed the ends to a nerve, the muscles of the frog preparation contracted. This suggested that the effect couldn't be due to animal electricity stored in the muscles (since they weren't part of the circuit) but must be due to some property of the bimetallic arc. Volta began to experiment with arcs made from tin, lead, or zinc in contact with silver or gold. When Volta stuck the ends of a bimetallic arc in his mouth, he tasted acid, an effect that had been announced by Johann Georg Sulzer in 1760.

Although Sulzer attributed the phenomenon to mechanical vibrations, Volta interpreted this perception as the electrical stimulation of his gustatory nerves. When he put the contacts of the arc against the surface of his eye, he saw light.[24]

Galvani responded to Volta both by demonstrating contractions using wire made from a single metal and by eliminating the wire in favor of a piece of the animal's own tissue, placed in such a way as to connect muscle and nerve. To Volta, these experiments were inconclusive because he believed that the nerve could be stimulated by mechanical or chemical means. Furthermore, Volta thought that contact between two tissues with different conductive properties might result in a charge, the same way that contact between two nonorganic conductors apparently did (as in the case of bimetallic arcs). But Galvani continued to attribute his findings to the presence of an innate animal electricity. In May 1795, he traveled to Sinigalia and Rimini so that he might experiment with electric fish. Placing a prepared frog on a live torpedo, he observed the expected contractions. When he then removed the frog from the fish, he found that he could induce further contractions by dangling it from the hind legs and briefly touching the exposed nerve with his finger. "I thought in this case electricity was communicated from the torpedo to the frog," Galvani wrote in an unpublished notebook, "and had charged the little Leyden jars which I supposed were there." He went on to speculate that the frog might even act as what we would call a rechargeable battery: that it could be possible to charge the frog with the torpedo, discharge the frog, and charge it again.[25]

Volta's thoughts about the frog preparations were running along different lines than those of Galvani, and by 1796 he was ready to abandon working with the animals altogether. He observed that the quantity of electricity that was required to stimulate muscular contractions in frog preparations was surprisingly small. Historian Marco Piccolino writes, "Having realized that the frog was more sensitive than his own sophisticated physical instruments, Volta started to view the frog as an extremely sensitive biological electroscope." Furthermore, while doing experiments on his tongue, Volta had noticed that the acid taste remained as long as the bimetallic arc was in his mouth. It appeared to him as if dissimilar metals in contact were somehow capable of generating electricity, like an electrical machine, rather than merely conducting it. Instead of continuing to work with frogs, Volta decided to find an alternative way to measure the tiny charges created by bimetallic arcs.[26]

In June 1796, Volta asked his instrument maker Giuseppe Re to create a doubler based on William Nicholson's design. This was a device that allowed minute electric charges to be accumulated to the point where they could be measured. Traditionally, small amounts of electricity were detected by the attraction or repulsion of lightweight bodies (familiar today as the "static cling" or "flyaway hair" of advertisements). In the electrometer, for example, a small charge registered by pushing apart two slips of gold leaf or a small pair of corks suspended on silver wires. Some charges, however, were too small to detect directly. In earlier work, Volta had used a device called an electrophorus, which allowed the experimenter to move a tiny charge from an electrostatic machine to a Leyden jar. But this was labor intensive because each charge had to be moved separately until enough of them added up to measure with an electrometer. William Nicholson's doubler automated this process, multiplying the charge each time a handle was cranked. (The engineer P. E. K. Donaldson, who has re-created charge multiplying devices, writes that "like so many other manual processes in the 18th century, it was ripe for deskilling by mechanization.") Working with the doubler, Volta became convinced that plates made of dissimilar metals could create electricity, but his experiments failed to convince fellow experimenters at the time.[27]

Galvani's experiments combining frog preparations with live torpedoes were published in 1797. That same year William Nicholson published a paper suggesting that the torpedo and other electric fish might be able to produce electric shocks by using a mechanism similar to Volta's electrophorus. By manipulating an electrophorus in a particular way, the device could be made to give any number of shocks. Nicholson conjectured that it should be possible to build a mechanism that retained its power to shock indefinitely. "I will not here describe the mechanical combinations which have occurred to me in meditating on this subject, but shall simply shew that the dimensions of the organs of the torpedo are such as by certain very possible motions, and the allowable supposition of conducting and non-conducting powers, may produce the effects we observe." In Nicholson's model of the action of strongly electric fish, large quantities of low-intensity electricity could be generated by pulling apart a lot of thin membranes, much as small charges were generated in his instruments by pulling apart thin discs of Muscovy talc.[28]

Volta biographer Giuliano Pancaldi argues that within a few months of reading the Nicholson paper, Volta had translated it into his own terms and built a working apparatus. He began by alternating disks of two dissimilar metals in a pile but failed to generate electricity this way. Eventually he discovered that by interposing wet paper disks between the metal ones he could generate an electrical current. The effect was enhanced by using a salt solution to wet the paper disks. In a letter read before the Royal Society on June 26, 1800, Volta announced the creation of an "*Organe électrique artificiel*" comparable to the natural electric organs of torpedoes, eels and other electric fish. Volta's model of the torpedo electric organ was different from both Cavendish's (a series of Leyden jars) and Nicholson's (a series of electrophores) because the earlier models relied on the alternation of conductors and nonconductors. The voltaic pile, however, was built entirely from conductors and yet could still deliver a shock. Volta's physical construction of the device as a column of thin discs was explicitly modeled on John Hunter's anatomical description of the columns that were to be found in the torpedo electric organ. Volta was eager to underline the similarity of the artificial electric organ to the natural one: "Is it not also active of itself without any previous charge, without the aid of any electricity excited by any of the means hitherto known? Does it not act incessantly, and without intermission? And, in the last place, is it not capable of giving every moment shocks of greater or less strength, according to circumstance—shocks which are renewed by each new touch, and which, when thus repeated or continued for a certain time, produce the same torpor in the limbs as is occasioned by the torpedo, &c?" "The circumstance that the battery (later depicted as a momentous turning point in the history of physics) was the outcome of a program that included the goal of 'imitating the electric fish,'" Pancaldi writes, is ". . . a cautionary tale against the sort of history of science that prevailed when the disciplinary boundaries imposed by twentieth-century developments oriented the work of historians."[29]

Volta's battery was taken up with enthusiasm. It could be made from readily available materials and was described in enough detail in his letter to be easily reproduced. William Nicholson and Anthony Carlisle, for example, built a voltaic pile (as it came to be known) from a stack of silver half-crowns interspersed with zinc discs and replicated Volta's experiments within a few months of seeing his letter. Furthermore, the battery promised to be fun to experiment with. Volta described a range of phenomena that could be elicited with the pile: perceptible sparks and shocks, and painful sensations if wired to the nose, eyelid, or forehead. If a person held a pile between the lips and touched it with their tongue, they would experience "a sensation of light in the eyes, a convulsion in the lips, and even in the tongue, and a painful prick at the tip of it, followed by a sensation of taste."[30]

Much like the earlier announcement of the Leyden jar, this sounded like something that could harm you if you weren't careful. Volta described connecting a pile of forty elements to wires that he then stuck in his ears. "I received a shock in the head, and some moments after . . . I began to hear a sound . . . ; it was a kind of crackling with shocks, as if some paste or tenacious matter had been boiling." Worried that this might be dangerous, Volta decided not to repeat the experiment, lest the "disagreeable sensation" be accompanied by a "shock in the brain." Who could resist building such a device? The historian Joost Mertens argues that it was precisely the fact that the battery was capable of being used for amusing public demonstrations that multiplied the number of nonexpert witnesses, helped establish metallic electricity as a fact, and made it clear that this part of physics was Volta's "jurisdiction."[31]

## A Vital Action

Alexander von Humboldt witnessed a demonstration of Galvani's frog experiments in Vienna in 1792 and began conducting electrical experiments of his own. Despite Volta's contention that the galvanic effect was an artifact of dissimilar conductors, Humboldt believed that Galvani's theory might lead to an understanding of vital forces. He systematically tested the sensitivity of about 300 species of plants and animals to electricity, including invertebrates, amphibians and reptiles, fish, birds, and mammals. The animals all responded, but even plants that were sensitive to physical contact, like *Mimosa pudica*, did not. During a vacation, Humboldt visited Volta while he was developing the voltaic pile. Humboldt came away with the idea of interspersing animal muscle tissue with metal plates, demonstrating that the force of the muscle contraction in a prepared frog depended both on the metals used in the pile and on the chemical and anatomical condition of the frog itself.

In a letter of 1796, Humboldt described "very strange experiments" that he had conducted, showing that frog's thighs that gave no response to zinc and gold would begin to contract if previously soaked in a potash solution. The effect even worked with homogenous metals instead of dissimilar ones. Eventually Humboldt managed to generate an electrical current from a pile composed entirely of layers of nerve and muscle tissue, eliminating the metal plates altogether. The "vitality" of the tissues used for this kind of experiment was crucial. Humboldt discovered that he could cut the crural nerve of a recently killed frog and separate the pieces by as much as two millimeters and muscular contractions would continue. If the animal had been dead for five minutes or more, however, any gap made in the nerve would stop the contractions.

Humboldt also used himself as an experimental subject, showing that electrical activity was much more intense if his skin was removed before applying the galvanic battery to an open wound. Humboldt's two volumes on galvanic experiments were published in 1797 and 1799, as he was making plans for a long voyage to South and Central America. In the first volume, he espoused the idea that muscles had an innate power of contraction, stimulated by the flow of galvanic fluid in the nerves of the living creature and by contact with metal or acid in the prepared frog. By the time that the second volume was published, Humboldt had retreated somewhat from his earlier conception of vital force, admitting that his observations might be explained by appealing to the dynamic balance of known physical forces. "Light, heat, electricity and the other components of the atmosphere, in which all creatures are bathed," he wrote, "operate upon them at every moment."[32]

In June 1799, Humboldt embarked on a five-year voyage that would take him from Venezuela to Mexico, accompanied by the botanist Aimé Bonpland. In Calabozo, Venezuela, amid the vast grasslands of the Llanos, they found a man named Carlos del Pozo who had devised a large electrical machine, a battery of Leyden jars, electrophori, and electrometers. It was a setup to rival the best that a natural philosopher in Europe might possess. Pozo was entirely self-taught. That he could build the electrical equipment by himself was testimony not only to his own ingenuity but also to the level of detail with which experiments were described in the literature of eighteenth-century natural philosophy. Pozo was delighted to meet the travelers, as his experiments up until their arrival had only been performed to the "astonishment and admiration" of "persons destitute of all information." He was also interested in the Leyden jar and electrometers that Humboldt had brought for physiological experimentation. Humboldt and Bonpland demonstrated the galvanic effect of touching bimetallic arcs to frog's nerves, as "the name of Galvani and Volta had not yet resounded in those vast solitudes."[33]

Besides their encounter with Pozo, Humboldt was excited to discover that there were electric eels in Calabozo, "animated electrical apparatuses." Up until that time he had been trying to procure some electric eels, but without success, as his money had little value in the interior. On the Caribbean coast of Cumana some Guayqueria Indians had brought him a small electric ray that was physically robust, but its electric shocks were very weak. Near Calabozo, however, the confluents of the Orinoco River were filled with electric eels. Humboldt and Bonpland went to the river's edge to watch Indians catch the fish for them. The electric eels could not be captured with nets because they buried themselves in the muddy river bottom. The traditional method—stunning the fish with "the roots of the piscidea erithyrna, jacquinia armillaris, and some species of phyllanthus"—would not work either, as they were effective precisely because they weakened the electric eels. The Indians told Humboldt and Bonpland that they would be fishing with horses instead. Driving about thirty wild horses and mules into the pool, they forced them to remain there, milling about and trampling the eel's habitat. The enraged eels swam to the surface to repeatedly shock the horses. "A contest between animals of so different an organization furnishes a very striking spectacle." Watching some of the stunned horses succumb to the assault and drown within minutes, Humboldt believed that the fish would eventually kill all of the horses. Gradually, however, the eels became exhausted and five large ones were taken with small harpoons attached to very dry cords. Humboldt carefully studied their anatomy, and warned against the temerity of exposing oneself to the first shocks of an irritated eel: "If by chance you receive a stroke before the fish is wounded, or wearied by a long pursuit, the pain and numbness are so violent, that it is impossible to describe the nature of the feeling they excite. I do not remember having ever received from the discharge of a large Leyden jar, a more dreadful shock, than that which I experienced by imprudently placing both my feet on a gymnotus just taken out of the water. I was affected the rest of the day with a violent pain in the knees, and in almost every joint."[34]

Humboldt noted that the skin of the electric eel was covered with a mucus that Volta had shown was twenty to thirty times more conductive than water. He observed that it was remarkable that none of the strongly electric fish yet discovered had scales and that strongly electric animals were found in water, which was an electrical conductor, but not in air. After extensive experimentation, Humboldt concluded that the electric eel was not a charged conductor, a battery, or an electromotive apparatus because "the action of the fish depend[ed] entirely on it's will" and could be directed toward external objects. Humboldt and Bonpland tried each holding one end of the animal and standing on moist ground. One felt a shock and the other did not, an effect that could be repeated when their hands were as close together on the eel's body as an inch. Sometimes one of them would handle the animal and not receive any shocks. Other natural philosophers had reported similar results. "The electric action of animals being a vital action," Humboldt wrote, "and subject to their will, it does not depend solely on their state of health and vigour."[35]

On his return to Europe, Humboldt decided to experiment with torpedoes. He and his friend and collaborator Joseph Louis Gay-Lussac managed to find some of the fish at Genoa but didn't have any instruments with which to test them. After trying to locate torpedoes in a few other places without any luck, they went to Naples, where they found that large and vigorous specimens were easy to obtain. Although the torpedoes were not as powerful as the electric eels, they were still fully capable of creating sensations the experimenters found highly aversive. Unlike the electric eel, which made no movement while discharging a shock, the torpedo flexed its pectoral fins convulsively before each stroke. As with the eel, touching the torpedo was not like touching a Leyden jar or galvanic battery: "it must be irritated that it may give its stroke; for this action depends upon the will of the animal, which in all probability does not always keep its electric organs charged: it charges them, however, with astonishing celerity, and is thus able to give a long series of shocks." Using an electrometer, Humboldt and Gay-Lussac tried to determine how charged the electric organs of the torpedo were, but without success. Humboldt also noted that "the least injury to the brain of this animal prevents its electrical action" and concluded that its nerves must be serving a primary function in the production of electricity.[36]

Returning from his journey to South America, Humboldt was taken with the *Naturphilosophie* of Friedrich W. J. Schelling, which seemed to accord well with his own view that the ties uniting geological and biological phenomena, and the relations between varied organisms, "are discovered only when we have acquired the habit of viewing the Globe as a great whole." In Schelling's formulation, all natural phenomena were due to the interplay of opposing attractive and repulsive forces: positive and negative electricity, north and south magnetic poles, chemical acids and bases, and so on. The goal of natural philosophy was to seek an underlying unity; the mind could only be said to understand nature "where it discovers the greatest simplicity of laws amid the greatest variety of phenomena, and the most stringent parsimony of means in the highest prodigality of effects." As the mind constructs an understanding of external forces, it has to do so by making use of its own resources, and thus for Schelling, "the system of nature [was] at one and the same time the system of our mind." The historian Robert J. Richards notes that it was exactly this aspect of *Naturphilosophie* that appealed to Humboldt upon his return to Europe. By traveling, by encountering raw, primitive nature, Humboldt had discovered himself. His "retrospective creation of his experiences in the jungles around the Orinoco—where he grew into that true self he became—those re-created experiences in his many descriptive volumes bear the mark of his engagement with Schelling."[37]

## Miraculous Fluids

Science captivated literate audiences of the late eighteenth century, the historian Robert Darnton shows, because it placed them in a world of "wonderful, invisible forces." In addition to gravity and electricity, there were "the miraculous gases of the Charlières and Montgolfières that astonished Europe by lifting men into the air for the first time in 1783." The Bavarian Academy of Science posed the problem of whether or not electricity and magnetism were physically analogous as its prize question for 1774–76. Competitors discussed a variety of topics ranging from the magnetic consequences of lightning to Schilling's claim of mutual attraction between electric eels and lodestones. In this context, another invisible fluid, such as that proposed by Franz Anton Mesmer in the 1770s, "seemed no more miraculous, and who could say that it was less real than the phlogiston that Lavoisier was attempting to banish from the universe, or the caloric he was apparently substituting for it, or the ether, the 'animal heat,' the 'inner mold,' the 'organic molecules,' the fire soul, and the other fictitious powers that one meets like ghosts inhabiting the dead treatises of . . . respectable eighteenth-century scientists." In the long run, animal magnetism proved to be a misstep rather than a step into a brave new world. But it took some time for that to become clear.[38]

A physician living in Vienna, Mesmer presented his doctoral dissertation on planetary influences on the human body in 1766. In it, he claimed that the movement of sun and moon produced atmospheric tides that altered the "gravity, elasticity, and pressure" of the air and thereby perturbed nervous fluid or animal spirits. Mesmer practiced an orthodox (for the time) form of medicine for many years, but his blistering, bleeding, and purgatives did not always effect the cures that he sought. One particularly recalcitrant case was a young woman named Franziska Österlin who had a raft of symptoms including blindness, earaches, toothaches, suffocation, inflammation of the viscera, vomiting, urine retention, melancholy delirium, and paralysis alternating with convulsions. Mesmer was at wit's end until he encountered a Jesuit priest named Maximillian Hell in 1774. Hell had previously given a heart-shaped steel magnet to a baroness who was cured of abdominal pains within a few days by applying it to herself. Hell suggested that Mesmer might try treating Österlin with magnets, too. So Mesmer did. He started with heart-shaped magnets attached to her chest and feet. She reported piercing pains and a burning sensation in her joints and pleaded with him to stop the treatment. Mesmer responded as any practitioner of heroic medicine would: by attaching more magnets to her person. Gradually her symptoms lessened until she was judged to be cured. As Österlin seemed to respond to magnets whenever she had a relapse, Mesmer suggested that she wear them prophylactically.[39]

Mesmer and Father Hell subsequently conducted a public dispute in an exchange of letters about who should receive credit for the idea of magnetic therapy. Mesmer claimed to have identified "animal gravitation" as a significant force in his doctoral dissertation, which his twentieth-century biographer Frank A. Pattie showed was actually plagiarized in large part from a 1704 book by an English physician. In fact, gravity in Mesmer's early thinking was an external force rather than an internal one. Mesmer believed that any change in this external force might damage the human body. As one example, he cited the case of a woman "whose features were regularly deflected to one side at the full moons, an effect very damaging to her good looks." It was only after his encounter with Father Hell that Mesmer began to think of this force as an internal magnetism rather than external gravitation. Mesmer quickly dispensed with the magnets, discovering that he could "magnetize" anything that he touched: men, dogs, metal, wood, stone, water, glass, silk, leather, paper, and bread. He could even fill bottles with his magnetic fluid or propel it eight or ten feet from his body by the strength of his will. Armed with this heretofore unknown force, Mesmer began a series of miraculous cures.[40]

The scientific academies of Europe ignored or dismissed Mesmer when he tried to correspond with them, finding his theory of animal magnetism to be wanting on several counts. Mesmer was publicly humiliated when he tried to demonstrate to Jan Ingenhousz that Miss Österlin would convulse in response to magnets, as Ingenhousz showed that she needed to believe that the objects were magnetic before she would respond. Ingenhousz denounced Mesmer as a fraud, leading to the public exchange of more angry letters. In 1777, Mesmer took on the treatment of eighteen-year-old Maria Theresa Paradis, a blind pianist who was financially supported by the Queen-Empress Maria Theresa and who had been (heroically) treated by the leading physicians of Vienna. Her poor eyes had failed to respond to the full gamut of leeches, blistering plasters, purgatives, diuretics, cauterization, and so on. Even discharging Leyden jars through her eyes thousands of times had failed to cure her blindness. Under Mesmer's care, she seemed to miraculously regain some degree of sight, and as a result her life took a turn for the worse. Her ability to cope with daily activities and her piano playing both declined precipitously, putting her royal pension in jeopardy. Other physicians questioned whether she could really see anything after all. Things came to a head with a ridiculous confrontation between Mesmer; Miss Österlin's sword-wielding and cursing father; her mother, who bashed the girl's head into a wall and then fainted; and the young woman herself, raging, vomiting, convulsing, and eventually losing her vision and suffering a complete relapse. Mesmer ineffectively stayed on the case for six months longer, until the resulting ignominy forced him to flee for Paris in 1778.[41]

The City of Light was more than willing to embrace sensation and novelty, and Mesmer quickly accumulated a clientele of wealthy and noble patients there. He developed methods to conduct his magnetic fluid to groups of people rather than treating them individually. One of his devices, the baquet, was a large wooden tub of "magnetized" water with angled metal rods protruding through the cover and extending around the perimeter. Patients could sit around the baquet at the same time, each touching one of the rods to an afflicted part. Patients were also instructed to hold hands with one another, or were looped with ropes, as Mesmer and his assistants passed their hands or metal wands over them to direct the flow of magnetic fluid. Each of these therapeutic innovations was firmly dependent upon his audience having had prior experience with the ritual and material paraphernalia of electrical demonstration. Despite his success in fashionable society, however, Mesmer still could not get scientific or medical academies to take him seriously. He managed to attract only one disciple with any professional standing, the physician Charles d'Eslon. D'Eslon wrote a book in support of mesmeric therapy and defended Mesmer to the Faculty of Medicine, a move that resulted in his own censure. When d'Eslon managed to establish a clinic of his own and begin successfully treating patients with animal magnetism, Mesmer was infuriated, believing that d'Eslon had betrayed him and stolen his techniques and ideas.[42]

Lay mesmeric societies proliferated throughout France, further enriching Mesmer and alarming the government of the ancien regime. In 1784, King Louis XVI established a royal commission to investigate mesmeric activities and to determine whether it existed or was useful, for "the animal magnetism may indeed exist without being useful, but it cannot be useful if it do not exist [*sic*]." Among its members were Benjamin Franklin, the chemist Lavoisier, and physician Joseph-Ignace Guillotine. Testing with an electrometer and nonmagnetic iron needle, the commissioners demonstrated that the baquet did not contain any electric or magnetic substance and thus had no physical capacity to create magnetic effects. (They were discounting Mesmer's claimed ability to magnetize bread, wood, water, dogs, and so on.) The welter of conflicting reports from attendees of group séances didn't seem worth sorting out, so the commissioners subjected themselves to treatments by d'Eslon. As was customary in electrical practice, they were using their own bodies as instruments. None felt a thing. After studying the effects of mesmeric treatment on various other patients, the committee observed that subjects who were incredulous tended not to receive any benefit. They concluded that the real cause of the effects attributed to animal magnetism was the imagination of the subjects. The commission's published report was widely distributed, making an ass of Mesmer by 1785—as that is how he was frequently caricatured in the popular press. Mesmer tried to continue his practice for a short while before abandoning Paris for a life of comparative obscurity.[43]

The possibility that "nervous fluid" might be electrical in nature expanded the domain of animal electricity from a few species of strongly electric fish to all animals, and it promised to unify natural philosophy across formerly separate realms. It also suggested that some forms of illness might be due to an imbalance of positive and negative electric charge that could be remedied by electrotherapy. Of course, some remedies (then as now) turned out to be bogus. Strongly electric fish continued to serve as experimental and therapeutic apparatus, joined now by great quantities of dismembered frogs and built into increasingly hybrid assemblages. The creation of the voltaic pile showed that it was possible for a purely inorganic device to generate electricity, but people and other animals would continue to remain in the circuits. This was especially true as electricity was turned toward probing the boundaries between life and death.

FOUR

# The Spark of Life

## Resuscitation

Unlike Mesmer's magnetic fluid, electricity remained miraculous. In 1774, a child named Sophia Greenhill fell from a window and was taken to Middlesex Hospital, apparently dead. The surgeons and apothecary could do nothing for her. After twenty minutes, a Mr. Squires of Soho was allowed to use electricity to try to revive her. He shocked various parts of her body without effect until he detected a small pulse in response to a shock to her thorax. In a few minutes, the girl began to breathe in a labored manner, eventually vomited and remained for several days in "a kind of stupor, occasioned by the depression of her cranium." She eventually recovered with further treatment. The Royal Humane Society reported the case and several subsequent ones where electricity was used to resuscitate victims of near drowning, lightning strike, asphyxiation, falls, and other mishaps. Within a few decades, some physicians would also recommend a galvanic shock applied to the neck in the region of the phrenic nerve as a means of distinguishing real from apparent death.[1]

The following year, the Danish physician and veterinarian Peter

Christian Abildgaard reported on some experiments that involved killing hens with an electric shock. In itself, this was not really news, as other people had been doing similar things since the birth of the Leyden phial. Abildgaard's innovation was to render the birds apparently lifeless with a powerful electric shock to the head and then to revive them with a second shock to the chest. He had been led to this unusual procedure by considering reports of people and animals killed by lightning. When their bodies were dissected, a clear cause of death was rarely identifiable. He could imagine three possible explanations: the air might become rarefied (as it did when using an air pump) and this could be the cause of death, or death could be due to ruptured blood vessels, or lightning might result in fainting followed by death. If it were the third case, as he suspected, then he reasoned that a victim of lightning might be revived with a second shock.

Abildgaard first tried to kill a three-month-old horse with a shock to the head from ten jars wired together, but "with no amount of effort could I kill [this] Herculean animal." He then succeeded in killing a hen with a shock to the head, but he could not revive it when he gave it further shocks to the head. It just lay there. Perhaps resuscitation was outside the realm of possibility? Abildgaard did not want to give up. He tried shocking the animal through the chest and spine, and it suddenly started. Setting it on the ground, he had the satisfaction of seeing the hen walking around quietly. He repeated the experiment several times, killing and reviving the hen until it was completely stunned and unable to walk, but it perked up within a couple of days "and even laid an egg." Concerned that the animal might have recovered without his intervention, Abildgaard tried shocking a cock and hen and leaving them. The next morning they were cold and dead, and no amount of shocking from the electrical machine could bring them back to life.[2]

Although vivisectionists had earlier discovered that electricity seemed to restore a bit of strength and vitality to their tortured experimental subjects, the idea that electricity might restore life to the recently dead was something new. Experimenters now tried resuscitating drowned fowl and suffocated frogs with shocks from an electrostatic machine. They found that a strong shock would kill a chick, a kid, or a lamb outright, whereas milder shocks simply paralyzed them. The muscles of animals that had been electrocuted were in rigor mortis and were not susceptible to electrical stimulation. A few held hope that electricity might reanimate the long dead. There were reports of electrically resuscitating a frog whose heart had been stopped an hour before. Alexander von Humboldt, too, was interested in the life-giving properties of electricity. Finding a dead linnet in his garden, he put a flake of zinc in its beak and wire of silver in its rectum and connected the two "armatures." "To my amazement," he wrote, "at the moment of contact the bird opened its eyes and raised itself on its feet by flapping its wings. It breathed anew for seven or eight minutes, and then expired quietly." Humboldt tried connecting his own mouth and anus with a bimetallic arc and saw vivid flashing, an effect that was first reported by Franz Karl Achard. These flashes reportedly became brighter and more colorful in darkened rooms, during storms or if one had an eye infection.[3]

Like Humboldt, any number of observers found that they would see flashes upon inserting galvanic probes into their eyes and mouths, between gum and upper lip, and up their noses and other orifices. The physician Richard Fowler, in addition to performing electrical experiments on frogs, worms, rabbits, cats, dogs, sheep, cows, and skates, stuck electrical probes in his own ears and felt his head jerk disagreeably. He wasn't sensible of sustaining any damage, but when he awoke the following morning his face and pillow were crusted with blood from one of his ears. He decided not to repeat that experiment, although he did try "insinuating a rod of silver, as far as possible, up [his own] nose." That was not necessarily safe, either. In 1805, Johann Wilhelm Ritter used the current from a voltaic pile made of twenty pairs of discs to experiment with his own Schneiderian membrane (lining the nasal cavity) and described the experience as "awful." He perceived a peculiar smell during the experiment and for some time afterwards. Electricity applied in one direction caused an acid smell and a loss of the ability to sneeze, followed by the smell of ammonia and a bought of sneezing when the circuit was broken. Current applied in the other direction reversed the order of the experiences, so that he smelled ammonia and sneezed first.[4]

The human body served as both instrument and brute matter for the would-be colonists of an electric world. The butchery of the French Revolution provided sentimental empiricists with an abundance of recently killed human flesh—at least for those natural philosophers who weren't themselves rendered into meat by the guillotine. "One minute before three, the axe fell on the Place de Grêve," one experimenter wrote, "and at 3.15 I already had the head in my hands and Mr. Nysten the body." The physician Pierre-Hubert Nysten did galvanic experiments on the cadavers of the executed, no doubt at considerable risk to his own safety, given the prevailing mood. In his *Nouvelles expériences galvanique* (1803) he compared his findings on humans with those he obtained from a variety of decapitated, strangled, and gassed dogs, guinea pigs, pigeons, frogs, and carp. Nysten discovered that the bulk of the human viscera was already insensitive to mechanical and electrical stimuli when he tested it within 40 to 60 minutes after death. The right ventricle was one exception. It could be excited for almost two hours, and electric shock made it susceptible to mechanical stimulation, too. Nysten was able to replicate his findings with the heart of a dog, showing that it became more excitable with repeated shocks, even after it had apparently lost its "*chaleur vitale*." Human voluntary muscles could be electrically stimulated for more than four hours, as could the right and left atria of the heart. Another researcher had the opportunity to experiment with the amputated leg of a boy, showing that it behaved just like a frog's: it contracted violently when he laid bare the crural nerve and stimulated it with a bimetallic arc. The boy's leg continued to respond until it was cold, 38 minutes after the amputation had occurred.[5]

Nysten was by no means the only natural philosopher to electrify the remains of recently killed people. Galvanic stimulation experiments were done with the bodies of decapitated criminals in Turin in 1802. But the most sensational experiments on human remains were begun that same year by Galvani's nephew and one-time pupil and assistant Giovanni Aldini, long a student of animal electricity. Aldini had already shown that a single metal, mercury, could cause contractions in a prepared frog if it was used to bridge nerve and muscle. He had tried using blood, bile, and urine, instead of saline, to moisten the layers between discs in a voltaic pile. He also reported that the muscles of a prepared frog would contract if its nerves were thrust into a wound cut in the muscle of another, living animal. Humboldt praised Aldini for his methodological ingenuity and his skilful experimentation.

Aldini took the lead in defending Galvani against Volta's attacks and upon his uncle's death in 1798 took the chair of physics at Bologna University. There he began experimenting with warm-blooded animals. He tried stimulating various parts of the brain of a decapitated ox to see which resulted in muscular responses: the cerebellum and corpus callosum gave the best results. Another physician who confirmed these findings concluded that the cerebellum of the live animal was itself a kind of electric battery that powered movement with galvanic fluid. Aldini also believed that the voltaic pile might be used for electrotherapy, so he tried applying it to human cadavers. "Convey an energetic fluid to the seat of all sensations," Aldini wrote, "distribute its force throughout the different parts of the nervous and muscular systems; produce, reanimate and, so to speak, control the vital forces: this is the object of my research."[6]

In Bologna in 1802, Aldini used a voltaic pile with 200 discs to stimulate the bodies of three criminals who had been decapitated about an hour earlier. For two hours he was able to elicit contractions in various voluntary muscles. The heart, by contrast, was not as responsive. With the assistance of another physician, Mondini, he was also able to stimulate the human corpus callosum, causing intense facial contractions—the head of "*la machine humaine*" behaved exactly like that of a decapitated ox. Aldini could even induce contractions on one side of the face by stimulating the contralateral hemisphere, although this effect was more-or-less ignored until it was rediscovered in the late nineteenth century. After his experiments, Aldini began a tour of Europe, putting on public demonstrations of animal electricity.

In Paris, he surprised the psychiatrist Phillipe Pinel by showing him the contractions that followed from galvanic shock; Aldini demonstrated on the body of an old woman at the Salpêtrière hospital who had died of "putrid fever" just a short time earlier. For various mental ailments, Aldini recommended connecting a voltaic pile to the hand and shaved, dampened head of the patient. He had obtained success with this method after first applying strong shocks to his own head and becoming insomniac for a few days as a result. Alas, the patients at the Salpêtrière were not as obliging as those in Bologna had been. Many were tied up and terrified. Since their hands weren't free, Aldini had to apply both leads from the battery to their heads, and the results were disappointing from his point of view (and no doubt much worse than that from theirs). After Aldini's visit, Pinel tried his hand at electrotherapy, but he left no record of his work. Presumably, he was no more successful than Aldini had been.[7]

On January 12, 1803, George Foster was sentenced to death at the Old Bailey for willfully drowning his wife, Jane, and infant daughter, Louisa, in the Paddington Canal. Foster was hanged a few days later and his body cut down and taken to a nearby house, where Aldini used it for galvanic experimentation. (The Italian had arrived in London only a week or two prior.) Several professional gentlemen were on hand to witness the proceedings. Applying electricity to the mouth and ear, Aldini was able to make the jaw quiver and one eye partially open. His subsequent electrifications of the cadaver made the arm raise, the fist clench, and the muscles of the legs contract. When Aldini touched the rectum with one of his electrodes, the convulsions of the body were apparently "so much increased as almost to give an appearance of reanimation." The *Newgate Calendar* says that "some of the uninformed bystanders thought that the wretched man was on the eve of being restored to life" and that the beadle of the Surgeons' Company "was so alarmed that he died of fright soon after his return home." The *Calendar* went on to reassure its readers that galvanism could be used to treat insanity, apoplexy, and other disorders and that in cases of drowning or suffocation it might restore breathing, "thereby rekindling the expiring spark of vitality." It was this very prospect of resuscitation that had led the Royal Humane Society to sponsor Aldini's visit to London.[8]

News of Aldini's experiments was received with excitement in some quarters and consternation in others. The *Times* of London reported Aldini's electrification of Foster's corpse on January 22, 1803. On February 4 they printed a satirical article that mentioned Sieur Robertson, who restored not only the lives, but also the former habits, of the dead with galvanism. Étienne-Gaspard Robertson was a Parisian showman who used a magic lantern to summon "Spectres, Phantoms and Ghosts" during public exhibitions that he called "Fantasmagoria." His advertised ability to animate the dead with galvanism was simply that: advertising. Another satire on March 4 suggested that a political accord between the English and French ambassadors to Naples was actually a result of Galvani electrifying them, so "that they arose by a simultaneous motion, stretched out their arms together by the next shock and, at the final vibration, shook each other by the hand violently during ten seconds." One on March 9 promised that Aldini would soon use electricity to "compel a fat Pluralist to put his head [*sic*] in his pocket, and divide its contents with the starving family of his Curate."[9]

Not all of the *Times* coverage was parodic. Aldini apparently performed a public demonstration of his experiments with decapitated oxen, and the "astonishing effects" were reported on February 15. Even the man in the street might half hope to profit from this new knowledge. When one of the characters in Thomas Skinner Surr's novel *A Winter in London* remarks on seeing a jockey at a lecture on galvanism, another replies, "Oh, my dear, he has killed so many racers, that he's half ruined in horse flesh; and so he expects, by learning Galvanism, to be able to bring his dead horses to life again." The grotesque powers of electricity to elide boundaries between people, beasts, and machines, between vital flesh and inert matter, were nicely captured by an anonymous satirist, who wrote a poem attacking Aldini and his ilk:[10]

For he ('tis told in public papers),
Can make dead people cut droll capers,
And shuffling off death's iron trammels,
To kick and hop like dancing camels.

To raise a dead dog he was able,
Though laid in quarters on a table,
And led him, yelping, round the town,
With two legs up, and two legs down.

. . . . . . . . . . . . . . . .

And this most comical magician
Will soon in public exhibition,
Perform a feat he's often boasted,
And animate a dead pig—roasted.

With powers of these metallic tractors,
He can revive dead malefactors;
And is re-animating daily,
Rogues that were hung once, at Old Bailey!

The finely tuned sensibility that had well served the sentimental empiricists of preceding generations was downgraded to mere sentimentality in the early nineteenth century. The historian Charlotte Sleigh notes that Aldini used the language of killing and reanimation to describe his feelings about his own work with human cadavers: a natural philosopher was "obliged to *suffocate* [his] *painful* sentiment in the desire to be of use to his *fellow kind* . . . the hope of pushing back the boundaries of human knowledge *reanimates* his courage and *silences Nature's voice*." In 1804 experiments, Aldini used electricity to stimulate the contraction of muscles around a cadaver's ribs, repeatedly forcing air through its mouth to snuff the flame of a candle placed before its head. (Talk about symbolic!) Pierre-Hubert Nysten continued to electrify the corpses of the recently deceased, while working at l'Hôpital de la Charité. Since muscles in rigor mortis did not respond to galvanic stimulation, he was able to use electric shocks to plot the course of rigidity as it descended through cadavers, from the face to neck, trunk, arms, and legs, an effect now known to forensic medicine as "Nysten's law."[11]

The Scottish chemist Andrew Ure performed galvanic experiments on the body of an executed murderer named Clydesdale in 1818, causing the face to grimace, the arms to raise, and point to spectators, the legs to kick assistants, and the like. For better effect, Clydesdale's cadaver was partially dissected to expose the spinal marrow, diaphragm, and sciatic and phrenic nerves directly to the battery leads. Ure's description of his experiments was dispassionately objective in places and intentionally shocking in others. When the face was electrified, he wrote, "Every muscle in his countenance was simultaneously thrown into fearful action; rage, horror, despair, anguish, and ghastly smiles, united their hideous expression in the murderer's face, surpassing far the wildest representations of a Fuseli or a Kean." The historian Iwan Rhys Morus argues that Ure's puppetlike treatment of the corpse should be read in the context of his subsequent advocacy for using unskilled women and children in the workplace. If natural economies were both galvanic and mechanical, they were susceptible to control and automation. Ure's work inspired similar experiments with the body of an executed murderer named Fallows in 1823.[12]

In 1818, the twenty-one-year-old Mary Wollstonecraft Shelley published *Frankenstein; or, the Modern Prometheus*. In later years, she recalled that the novel had been inspired by a dream that she had while visiting Lord Byron in Geneva with Percy Shelley in 1816: "I saw the pale student of unhallowed arts kneeling beside the thing he had put together; I saw the hideous phantasm of a man stretched out; and then, on the working of some powerful engine, shew signs of life, and stir with an uneasy, half-vital motion. Frightful must it be; for supremely frightful would be the effect of any human endeavour to mock the stupendous mechanism of the Creator of the world." Scholars have suggested that Shelley may have also been influenced, consciously or not, by another, heartbreaking incident she recorded in her journal. At seventeen, she had given birth to her first child, a baby girl who died two weeks later. Shelley wrote, "Dream that my little baby came to life again—that it had only been cold & that we rubbed it by the fire & it lived—I awake & find no baby."[13]

*Frankenstein* soon became emblematic of the monstrosity that might result when natural philosophers meddled with life and death. "Who shall conceive the horrors of my secret toil," Shelley's protagonist Victor Frankenstein lamented, "as I dabbled among the unhallowed damps of the grave, or tortured the living animal to animate the lifeless clay?" In fact, the toil of contemporary Frankensteins like Giovanni Aldini and Andrew Ure was not secret at all but proudly on display. In the year between Shelley's dream and the publication of her novel, the *Naturphilosoph* and physician Karl August Weinhold reported on a series of experiments that he had conducted on life and electricity. He had made some galvanic observations on decapitated human heads, but a Prussian law of 1804 made that practice illegal when it became too popular. So when Weinhold decided to create "an artificial brain and spinal cord" from a silver-zinc amalgam, he turned to lively and playful kittens instead.

> I removed the cerebrum and cerebellum, as well as the spinal cord, with a small spoon through an opening at the back of the head. After this, the animal lost all life, all sensory functions, voluntary muscle movement, and eventually its pulse. Afterward, I filled both cavities

> with the aforementioned amalgam. For almost 20 minutes, the animal got into such a life-tension that it raised its head, opened its eyes, stared for a time, tried to get into a crawling position, sank down again several times, yet finally got up with obvious effort, hopped around, and then sank down exhausted. The heartbeat and the pulse, as well as the circulation, were quite active during these observations and continued after I opened the chest and abdominal cavities 15 minutes later.

Weinhold was apparently detested by many of his peers, and, one gets the impression, by many of the people who have written about him since. He was reported to be an unpleasant beardless man with a high voice, small head, and long arms and legs. An outspoken proponent of infibulation for impoverished male youth, upon his death his own genitals were discovered to be deformed. Feminist scholars like Anne K. Mellor have read *Frankenstein* as "first and foremost a book about what happens when a man tries to procreate without a woman." The same kind of analysis may usefully be extended beyond the realm of fiction, of course.[14]

We now have defibrillators to restart hearts, pacemakers that regulate their activity, implantable microelectrode arrays connecting electronic circuits to living neural tissue, cochlear implants, and any number of other electric and electronic systems that partake of both body and machine. Many of these technologies would not exist if it had not been for the variety of grisly experiments undertaken in the long nineteenth century. Although these are usually read as episodes in the history of medicine and the history of technology, we can see them also as depictions of particularly unusual animal behavior. We humans are unique in our willingness to treat just about anything as *apparatus*, including ourselves, one another, human body parts, other animals, animal body parts, inanimate objects, and hybrids of some or all of the above. Many of our techniques have become more subtle than spooning out kitten brains to replace with galvanic amalgam, but they are no less invasive or hybridizing. Our proclivities in this regard are a large part of what makes us human. We are, the philosopher Andy Clark argues, "creatures whose minds are special precisely because they are tailor-made for multiple mergers and coalitions." "Tools-R-Us," he quips, "and always have been."[15]

## Romantic Electrochemistry

The voltaic pile provided experimenters with a steady electric current, and this, in turn, gave them new means for disassembling matter into its constituents. Chemistry began to emerge as a separate discipline in the mid-eighteenth century. The historian Thomas Hankins notes that the impetus for this was the recognition that "air," a traditional Aristotelian element, was actually a state of matter and that atmospheric air was comprised of different chemicals in a "vaporous" state. Elemental "fire" (including electricity) played a crucial part in the chemical analysis of air while resisting parallel efforts at rationalization.

In the 1770s, Joseph Priestley showed that exposing both common air and "dephlogisticated" air to electrical discharge made them less able to support the burning of a candle flame or the breathing of a living animal such as a mouse. Phlogiston, in some chemical theories of the time, was a firelike substance thought to be released as flame during combustion and to have something to do with respiration. In this view, the substance remaining after combustion was the true, "dephlogisticated" element. For example, when metals or minerals were heated in the process known as calcination, the phlogiston was driven off to create the true element, a calx. Under different conditions, fire could also be used to "revivify" metals from their calxes. Since electricity served as well as common fire in these metallurgical processes and in the combustion of gases, electric fire and phlogiston had come to be identified with one another during the second half of the eighteenth century.[16]

Electrochemistry emerged in the context of experiments on air and (electrical) fire. Henry Cavendish had begun his research in natural philosophy with a paper on the chemistry of arsenic. His earliest work exhibited the same tendency to treat his own perceptions in an instrumental fashion as his later work on the artificial torpedo did. Cavendish's biographers write: "In this chemical work, Cavendish's senses were fully engaged, and he described his sensations with a discriminating vocabulary. With colors, he made the most distinctions: milky, cloudy, yellow, pale straw, reddish yellow, pale madeira, red, reddish brown, dirty red, green, bluish green, pearl color, blue, and transparent, turgid, and muddy. By smell, he distinguished between the various acids and their products. He observed the degree of heat, the strength of effervescence, the speed of dissolution, the shape and size of crystals. He observed textures: dry, hard, thin jelly, gluey, thick, stiff mud, lump. No poet paid greater attention to his sensations than Cavendish did to his."[17]

After publishing the torpedo paper, Henry Cavendish did some further work on the electrical conductivity of solutions of salts, acids, and gases in water. In unpublished notes he compared the conductivity of each to a standard salt solution, arriving at values that James Clerk Maxwell would find were as accurate as those that could be obtained with the best instruments a hundred years later. Cavendish did not experiment with electricity again until briefly in the 1780s while studying air. Following observations made by Priestley and others, Cavendish showed that when various gases were combined in the right proportions and ignited with an electric spark, the explosion was followed by the condensation of liquid on the inside of the vessel. Depending on the gases used, this liquid was either pure water or dilute nitrous acid. Cavendish and Priestley understood these results in terms of relationships between phlogiston, dephlogisticated air, phlogisticated air, and water. To the French chemist Antoine Lavoisier, however, they showed that liquid water is actually a chemical compound made from two gases that he called oxygen (formerly dephlogisticated air) and nitrogen (formerly phlogisticated air). Cavendish's experiment provided one of the crucial clues that Lavoisier needed to reform chemistry.[18]

By the 1790s in both England and France, combustion was understood by most natural philosophers in terms of Lavoisier's oxygen instead of the older phlogiston. Lavoisier also introduced the idea that the phenomenon of heat was due to the transfer of a subtle fluid known as "caloric" from warm bodies to cooler ones. Lavoisier's supporters argued that electricity could decompose water into hydrogen and oxygen by serving as a source of caloric. Electricity thus continued to be associated with fire, even as Lavoisier and his followers extirpated phlogiston from chemical explanation. When Volta sent a letter about his artificial electric organ to the Royal Society in 1800, a copy was given to Anthony Carlisle and William Nicholson, who had recently developed his own model of the action of strongly electric fish. Nicholson and Carlisle built a voltaic pile and began experimenting with it. They agreed with Volta that galvanism was clearly electrical in nature but disagreed with his claim that it was due to contact between dissimilar conductors. Instead they favored a chemical explanation for several reasons, including the fact that the operation of the pile produced an acid. They observed that a droplet of water placed at the junction of the iron wire and the silver disc on the top of a voltaic pile developed bubbles of gas that smelled of hydrogen. When Nicholson and Carlisle put a tube of water into a voltaic circuit, they found that water could be galvanically decomposed into hydrogen and oxygen. Their discovery was heralded as promising to "throw light on several phenomena of the Animal Economy, as well as Chemistry and Electricity."[19]

Nicholson and Carlisle's results were widely discussed throughout Europe, where they served to consolidate previous observations. In the early 1790s, Giovanni Fabbroni had argued that galvanism could be interpreted chemically. The paper he read in Florence before the Accademia dei Georgofili in 1793 wasn't published until 1801, however, and French and English translations of his work had to wait until 1799 (including one that appeared in William Nicholson's *Journal*). Fabbroni interpreted the acid and alkaline tastes generated by a bimetallic arc to be due to an oxidation reaction that crucially involved the water in saliva. If the tongue was dried before touching it with dissimilar metals, the effect was much weaker. When a bimetallic arc was placed in water, the oxidation reaction could be stopped by pouring oil or mercury onto the surface, presumably because the supply of air was cut off. If the surface of the water was covered with a film of a material rich in oxygen, like an oxide, the reaction continued. In 1796, an English doctor wrote to Humboldt to tell him that water could be decomposed with galvanism, an effect Humboldt replicated, collecting bubbles of a gas that he could not identify (hydrogen). Johann Wilhelm Ritter presented similar arguments a few years later. Ritter had also observed that when Volta arranged metals into a series based on their properties for contact electricity, the metals were also ordered by their affinity for oxygen. All of these findings took on new significance after 1800, as the voltaic pile was gradually transformed from an experimental demonstration in its own right into an instrument that could be used for electrochemical research.[20]

## Davy

Humphry Davy played a central role in the conceptual transformation of current electricity and its mobilization in powerful new technologies. In 1797, at the age of eighteen, he began studying Lavoisier's *Traité élementaire de chimie* (1789) together with William Nicholson's *Dictionary of Chemistry*. Two years later, Davy published his first work, an "Essay on Heat and Light." In it, he claimed that light consists of material particles that are bound with gaseous oxygen. During combustion these particles are released, which is why fire gives off light. He also argued that since ice could be melted with friction alone, Lavoisier's caloric must not exist. Davy knew that oxygen was constantly consumed by animals during respiration and that Priestley and Ingenhousz had shown that it was created by plants. What he really wanted to do was to find the link between light and life. Davy's thesis was that oxygen is broken down to provide light to the brain, where it is transmitted to the nerves as a gas or subtle fluid. His biographer David Knight writes that "because electricity stimulated the nerves, [Davy] 'concluded the nervous fluid to be the electric aura'; sensations or ideas are motions of the nervous ether or light; and the 'irritability' of living matter is a consequence of contained light, which is essential to perceptive existence." Davy went on to speculate that all physical and chemical laws would be "considered as subservient to one grand end, *perception*."[21]

Volta's letter on the voltaic pile—to use a metaphor that was relatively new at the time—galvanized the community of natural philosophers. Joseph Priestley reflected that chemistry and the physics of light and color seemed to provide an avenue by which the internal structure of things might be known; natural philosophy would no longer be limited to an exploration of sensible properties. The ability to penetrate to the heart of nature was just what a young Romantic like Humphry Davy was seeking. He had been experimenting on himself with nitrous oxide and digitalis and writing poetry and detailed descriptions of his sensations under anesthesia, and he had become close friends with Samuel Taylor Coleridge. Reading Volta's paper, Davy found a new direction for his life's work in electrochemistry.

Davy's friend Thomas Beddoes made a voltaic pile with 110 pairs of discs. When Davy put pure water between the layers, the device failed to work. He concluded that the liquid between the conducting disks must be something that was capable of oxidizing the zinc and that the stronger this chemical reaction was, the greater the electric shock would be. He wasn't sure of the mechanism, but he was sure that electricity was generated by a chemical reaction and, in turn, that a chemical reaction could be caused by electricity, which is what Nicholson and Carlisle had demonstrated. By 1800, Davy was already sending papers on his experiments to Nicholson's *Journal*. These papers show that Davy was concerned not only with the chemical analysis for which he would later become famous but also with the parts and tissues of dead and living bodies, including his own. The following year, Davy described making a battery with a single disc of tin or zinc used to separate two fluids, one that oxidized the metal and one that could not. He also showed that batteries could be made of discs of zinc and charcoal and that using concentrated nitric acid instead of saline or dilute sulphuric acid made a cell that was more efficient, by about an order of magnitude. He was elected as a Fellow of the Royal Society in 1803.[22]

Metals, even dense ones like gold, contract when they cool. To Newton, and his followers like Priestley, this suggested that solid matter might consist of tiny particles that interacted with one another through forces. Newton had provided an analysis of gravity in these terms, but chemical forces must be more selective: all bodies are susceptible to gravitation, but only some interact chemically. The problem of chemical "affinity" thus came to the fore at the turn of the nineteenth century, conceptualized as analogous to the sympathies and antipathies that people felt for one another. Davy knew of Schelling's *Naturphilosophie*, which gave primacy to the interplay of polar forces—he was probably exposed to it by Coleridge—and he was familiar with the work of *Naturphilosophs* like Ritter and Hans Christian Ørsted, although it was too metaphysical for his taste. Like Newton, Davy preferred his matter "inanimate, brute and inert." Like Cavendish, Davy believed that atmospheric electricity, machine-generated electricity, animal electricity, and galvanism were all manifestations of the same phenomenon.

But was electricity a single fluid? Or two, positive and negative? If the latter, then the Nicholson and Carlisle demonstration might be interpreted as showing that oxygen and hydrogen were actually compounds made from elemental water and positive and negative electrical fluids. To eliminate that possibility, Davy spent years refining the basic experiment, replacing chemically active glass vessels with agate ones, using tubes made of gold, distilling his water in a silver still, and so on, removing all potential sources of chemical activity. He confirmed that electricity decomposed water into two elements, oxygen and hydrogen, in the correct proportions. He also showed that the chemical properties of metals were not simply attributes of the material but depended on their electrical states, too: silver was reactive if it had a positive charge, while zinc was inert if negatively charged. Davy predicted that natural electricity would be found "immediately and importantly connected with the order and oeconomy of nature; and investigations on this subject [could] hardly fail to enlighten our philosophical systems of the earth; and may possibly place new powers within our reach."[23]

It didn't take Davy long to make good on his prediction. Earlier he had remarked that galvanism "enabled men to produce from combinations of dead matter effects which were formerly occasioned only by animal organs." Now he would use dead matter to produce effects that no animal could . . . not even the strongly electric fish. Davy's experiments before 1808 were conducted at the Royal Institution with a battery made of one hundred double plates of copper and zinc, each six inches square. Toward the end of the decade, however, he was arguably one of the most important natural philosophers in the world, routinely compared to Newton, honored not only by his countrymen but also by the French. Staying ahead of rival French chemists required the best equipment. Proprietors and subscribers at the Royal Institution collected more than £1,000 to build a 2,000-cell "Great Battery" for Davy in the cellar. Its maximum power was about as much as that generated by a car battery when the engine is started on a cold day. In 1809 or 1810, Davy connected two rods of charcoal, each about an inch long and a sixth of an inch in diameter, to the leads of the Great Battery and brought them close to one another. A spark jumped the gap, igniting the ends of the rods to "whiteness." As Davy drew the rods apart to a distance of about four inches, the electric arc continued to bridge the space, filling the cellar with a brilliant light. Materials that he placed into the arc—diamond, sapphire, quartz, magnesia, lime, platina, plumbago, charcoal—fused, melted, or were vaporized. As the century progressed, electric lighting would perturb the circadian rhythms of ever more people, animals, plants, and microbes.[24]

The historian Giuliano Pancaldi shows that the voltaic battery was a "hybrid object" in the sense that it raised several interpretive puzzles. It straddled boundaries between animals and non-animal things, between animal electricity (galvanism) and common electricity, between chemistry and physics, and between natural entities and human-created artifacts. Although many subsequent biographers of Volta and Davy have treated the animals in their stories as superfluous to the physics and chemistry, we know that this was not the case for the men themselves. Davy's earliest activities were directed toward a unified understanding of life, perception, and physical forces. From 1808, he was a prominent member of the Animal Chemistry Club, a group that met to discuss physiology and whose members shared the premise that life could not be reduced to chemistry alone. Davy was also an avid angler his whole life. He fished to relax and shared the pastime with other chemists like William Hyde Wollaston.

Having read the papers of John Walsh, John Hunter, and Henry Cavendish, Humphry Davy also believed that study of the electricity of strongly electric fish would shed more light on electrochemistry. Visiting the Mediterranean coast in 1814 and 1815, Davy obtained living torpedoes to experiment with. He tried directing their shocks through silver wires into water to galvanically decompose it, without success. Changing the apparatus to use chemical conductors such as solutions of sulphuric acid and potassium hydroxide did not help. Davy spent some time with Volta in the summer of 1815 and mentioned to him that the torpedo did not seem to operate in the same manner as a voltaic pile. Volta showed Davy "another form of his instrument, which appeared to him to fulfill the conditions of the organ of a torpedo," a pile made with honey between the discs, "which required a certain time to become charged, and which did not decompose water, though when it was charged it communicated weak shocks."[25]

## Electricity and Magnetism

The connections between electricity, magnetism, and life remained to be elucidated. Strongly electric fish turned out not to be attracted to or incapacitated by lodestones, and Mesmer's debacle made animal magnetism appear to be a considerably less promising idea than animal electricity. Nevertheless, magnets could attract and repel at a distance. Building on Kant's *Metaphysical Foundations of Natural Science* (1786) and Schelling's *Naturphilosophie* (1797), the best way to explain action at a distance was to posit the workings of a fundamental force. In this view, electricity and magnetism *must* be closely related; it only remained for natural philosophers to figure out how.

Johann Wilhelm Ritter did some unpublished experiments with frogs in the 1790s, trying to clarify the effects of magnetism and galvanism on living organisms. Touching a frog with two pieces of iron or a chain made of iron and steel had no effect. When Ritter substituted a magnet or magnetized iron for one of the pieces of metal, the frog twitched. In 1798, Humboldt confirmed Ritter's results but discovered further puzzles. A pair of strong magnets stimulated contractions in the frog when placed with the same poles together but not when the opposite poles were touching. Even though he could not explain the mechanism, Humboldt was willing to speculate that magnets must have an effect on nutrition, fluid motion, and vitality when placed near living tissue. Ritter's own book attempting to demonstrate that galvanism accompanied the vital processes of animals was published the same year. With the introduction of the voltaic pile in 1800, it became clear to Ritter, Humboldt, and like-minded natural philosophers that studies of chemical reactions, electricity, magnetism, and the phenomenon of life could shed light on one another.[26]

Attempts to store magnetic force in a device analogous to the voltaic pile failed, but experimenters soon found another useful mechanism for electrical storage. While Nicholas Gautherot was using a voltaic pile to decompose salt water in 1801, he found that the silver or platinum wires used for battery leads could elicit contractions in a frog's leg even after they had been detached from the pile. Ritter observed a similar phenomenon two years later while using gold leads, and he replicated the effect with iron, brass, and bismuth wires. Ritter realized that the detached battery leads were serving as a storage device for electricity in their own right. Perhaps recalling Galvani's demonstration that a frog preparation could be charged and recharged by touching it to a live torpedo, Ritter took the further step of creating a "secondary" pile that could be charged and recharged by attaching it to a voltaic pile. In addition to the voltaic pile—for which Ritter coined the term "battery"—there was now another device that could provide experimenters with current electricity, what we now call a secondary, or rechargeable battery.[27]

Ritter speculated that every instance of electrical attraction and repulsion might also be magnetic. He believed that cats exhibited electrical polarity, and he wondered if that was true of all animals. Perhaps they were magnetized from head to tail, too. Perhaps the electricity could be removed if one ruffled their fur in the wrong direction. Maybe plants were electrically and magnetically polarized as well. The patterns he imagined extended beneath the skin: heads and brains were electrically negative, tails were electrically positive, and the nerves spread themselves from the spinal cord as if electrically repulsed. In his notes, he compared these neural arrangements to Lichtenberg figures. Here he was referring to the demonstration that Georg Christoph Lichtenberg had pioneered: fine dust, usually Lycopodium spores, spread on the plate of an electrophorus formed into fernlike patterns when a spark was discharged from the device. Although Ritter believed that the human nervous system was an exquisitely sensitive detector for fundamental forces—after all, he had done much painful auto-experimentation with electricity—he tried using Lichtenberg figures as instruments that could register electrical activity without human intervention. He had no doubt that the forces at play were ultimately physical, even though they were bound up with vital processes. He saw the patterns made by physical forces in the Lichtenberg apparatus as the pictorial language of the "Book of Nature" itself, "its hieroglyph, its character, its script."[28]

To Ritter, the analogy between the inorganic and the organic ran deep. A galvanic battery operated only as long as it had positive and negative electricity. He thought that sick animals also "suffer[ed] changes and weaknesses in their electricity." Sick cats were not electric, and "sick electric fish give much weaker charges. Thus a sick animal, a sick human, is like a used up battery, and a battery which has become weak is a sick one." Ritter imagined that experiments with strongly electric fish could prove decisive in clarifying the relationships between galvanism and animal magnetism, if he were able to chain together enough fresh frog legs and frog nerves to discriminate between effects due to each force. He thought there might even be worms and insects (such as the jumping beetle) which, if tested, could cause contractions in a prepared frog.

Sound, light, electricity, heat, and magnetism must all be transmitted by waves. Ritter's friend Hans Christian Ørsted hypothesized that our ability to sense these forces depended on the transmission of weak electric waves to eye or ear and then to the nerves. Ritter agreed with him. Furthermore, since light, heat, and sound could be reflected, Ritter imagined that it would soon be possible to reflect electricity and magnetism with mirror-like devices. If these forces could be manipulated in such a way, what instruments might be devised?[29]

Strongly electric fish continued to serve as both experimental subjects and a source of inspiration for new theories. Luigi Rolando, working on the Mediterranean island of Sardinia in the first decade of the nineteenth century, believed that the brain secreted a galvanic fluid that was conducted through the nerves and stimulated the muscles to action. In his account, the cerebellum was comparable to the electric organ of the torpedo, and served a similar function. For Rolando's contemporary, the British chemist William Hyde Wollaston, Humphry Davy's work in electrochemical separation and transfer suggested that "animal secretions were affected by the agency of a similar electric power" and the electric organs of the torpedo and *Gymnotus* served as proof of the existence of animal electricity that was internally generated. The *Naturphilosoph* G. R. Treviranus speculated that "perhaps it is the same power that through the [electric] organs produces the electric shocks, that is the immediate cause of the contraction [of muscle fibres]."[30]

In 1812 and 1816, John T. Todd, the surgeon of the Royal Navy vessel *Lion*, had the opportunity to experiment with some torpedoes caught at the Cape of Good Hope while the sailors were fishing for food. He compared their electrical organs with the anatomical studies of the torpedo published by John Hunter, noting that the largest proportion of nerves went to the creature's electrical organs. The fish that he observed and handled were small; he felt their shock as high as his elbow but never above his shoulder. There was no pattern to their electrical discharges. Sometimes they shocked him immediately and repeatedly, sometimes only sporadically. Occasionally they simply struggled to get out of his grasp without shocking him at all.

Todd took a pair of torpedoes and placed them in separate buckets of sea water, irritating one repeatedly while leaving the other undisturbed. The first soon died. He repeated that experiment several times with similar results. He tried vivisection, separating the nerves to the electrical organs of a lively torpedo and then leaving it to rest for a few hours. When he disturbed it, it could not shock him, and it lived about as long as the torpedoes who were left undisturbed in his previous experiments. He compared the results when he irritated an intact torpedo and one whose electric organ nerves had been divided: the ones who were able to discharge shocks died much sooner than the ones he had vivisected. He concluded that the torpedo's electric discharge was a vital action under voluntary control and that removing the fish's ability to shock might extend its life and make it more "vivacious," an effect he compared to castration in other animals. In subsequent vivisection experiments at La Rochelle, Todd observed that the electric organ was stimulated by cutting it with a wet scalpel.[31]

Meanwhile, Ørsted had been mulling over the relationship between electricity and magnetism for more than a decade when he hit on the idea of putting a compass near a wire that was carrying an electric current. In a public lecture on magnetism that he gave in the winter of 1819, he demonstrated that the needle deflected slightly as a result of the current. His audience, apparently, was not impressed. Ørsted conducted a series of further experiments with more powerful apparatus in the presence of other natural philosophers. He showed that the "conflict of electricity" that resulted in the wire and in the surrounding space when the positive and negative poles of a battery were connected, passed to the compass needle through nonmagnetic materials such as glass, metal, wood, stone, and ceramics. The particles of magnetic bodies, by contrast, "resist[ed] the passage of this conflict." Using the inclination and declination of the compass needle, Ørsted determined that the electric conflict "perform[ed] circles" around the axis of the current-carrying wire. Since he believed that heat and light were also due to the conflict of electricities, he predicted that the effect that he had observed would be used to explain the polarization of light. He also suggested that a magnetic pole should be able to deflect an electrical current in turn.[32]

Ørsted's work was published in 1820 and soon replicated in Paris. The news spread from there to England, where Humphry Davy and Michael Faraday began experimenting with the effect. They confirmed Ørsted's hunch that an electric current could be deflected by a magnet, and the following summer Faraday showed that a current-carrying wire would rotate around a magnetic core. Electricity, in other words, could be harnessed to do mechanical work, although it would take about a half century to make commercially viable motors. In our world, turbines driven by steam, water, wind, and combustion convert mechanical motion into electricity. This is distributed through networks, stored in batteries, and then converted back into mechanical work by an enormous variety of motors ranging in size from molecular to massive. We use these electric motors to move everything from minute drug doses to cruise ships.

Ørsted's discovery also deflected the research program of physicist André-Marie Ampère, who set to developing a new field of electrodynamics. In the 1780s, Charles Coulomb had derived an inverse square law to describe the attraction between electric particles with opposite charges and the repulsion between those with similar charges. To measure the forces that charged particles exerted on one another, Coulomb devised a "torsion balance." This consisted of an insulated bar with a metal ball at one end, suspended from a silk fiber. As another charged ball was brought near the ball on the bar, the bar would rotate toward or away from it (depending on the charge), thus twisting or untwisting the silk thread. Since Coulomb could measure the amount of effort required to twist the fiber through a given angle, he could measure the forces the charged balls exerted on one another. The charges were stationary in Coulomb's work (although the charged objects were not), so Ampère referred to it as "electrostatics." In his own electrodynamics, Ampère made clear the distinction between static and dynamic electricity, derived a law that described the interaction of two electric currents, showed that electric and magnetic fluids were one and the same, and described the magnet as merely a bundle of electric currents. His work combined the emphasis on advanced mathematics that characterized French physics with a thoroughgoing empiricism that was the hallmark of British researchers from Cavendish to Faraday.[33]

By the late eighteenth century, most observers agreed that the spark that distinguished the quick from the dead was electrical. Electrical fluids could suspend or take life, could resuscitate those near death, could—perhaps!—restore life if conditions were just right. It was a short, but fraught, step to the possibility that electricity might even be used to bestow life on inanimate matter.

Marie-François Xavier Bichat studied philosophy before turning his attention to pathology, and a certain degree of vitalism pervaded his work. He defined life as "the totality of those functions which resist death" and then set about systematically determining the conditions under which the vital functions stopped resisting: death of the heart, brain death, injection of air into the veins, suffocation, poisoning, and so on. In his work, Bichat emphasized experimental control and replication, and drew on the techniques of anatomy, vivisection and autopsy. He famously showed that organs are built up from the composition of tissues—he identified twenty-one different kinds—much the same way that compounds were made from elements in Lavoisier's chemistry, and, in so doing, he put physiology on a path toward the search for ever-smaller building blocks from which living things could be constructed. Bichat did many galvanic experiments with frogs; in 1789, he also had the opportunity of doing extensive galvanic experimentation with freshly guillotined human cadavers, working his way through hundreds by some accounts.[34]

Bichat, alas, was the victim of an early death. French physiology passed into the hands of men like François Magendie, who was a youth during the French Revolution. While rejecting Bichat's vitalism, Magendie adopted his methods wholesale, systematically vivisecting and killing one dog after another to demonstrate the action of poisons like strychnine, the mechanism of vomiting, and the function of the nerves and spinal cord in controlling sensation and motion. This last was one of his most enduring contributions: if one nerve was severed, an animal could still feel things but was immobilized, whereas cutting another nerve allowed the animal to move but removed its capacity to feel things. The heartlessness and occasional levity that Magendie brought to public vivisection was enough to outrage many Britons, making him one of the main targets of an emerging antivivisectionist movement.

Then, as now, mammals elicited more compassion than our cold-blooded relatives. In 1830, Leopoldo Nobili remarked, "For thirty years now frogs have been tormented with electricity, and the phenomena they exhibit still continue to excite the astonishment of an attentive observer." Such experiments had become the province of any amateur who was curious enough to try them. The "Electro-galvanism" article of the *London Encyclopaedia* of 1829, for example, carefully described dozens of demonstrations that could be done with a bit of electricity and intact, skinned, pithed, decapitated, dismembered, disemboweled, or desanguinated frogs. The human instrument still played a crucial role in these studies. One experiment involved removing a frog's leg and coating the nerve in zinc. The experimenter was then supposed to coat his tongue in silver and touch the nerve to it. This would result in the contraction of the frog's leg, but the experimenter would not taste anything. Coating the tongue in zinc and the nerve in silver, by contrast, would cause the frog's leg to remain immobile when nerve was touched to tongue, but a peculiar taste would fill his mouth.[35]

The connection between electricity and life was still far from being resolved. For some, like the colorful mercenary Colonel Francis Maceroni, as above, so below. Having witnessed underwater lightning storms and electrical discharges that ascended from sea to sky, and seen volleys of lightning accompany a volcanic eruption at Vesuvius, he argued that electrical processes of secretion, growth, and decomposition were necessary to both inorganic and organic matter at every scale. Any attempt to separate the two realms was bound to be futile. Maceroni saw analogies between the physiology of organisms and the physiology he attributed to the earth; between the atmospheric layers of air, cloud, water, and ground and the alternating layers of a voltaic battery; and between geological strata and organic tissues. Animal nervous systems, "evidently the seats and conductors of vital action," were clearly "real electrical machines; similar in principle, as they are similar in substance and in structure, to the electrical discharging-apparatus of the gymnotus and torpedo." For Maceroni, all vital phenomena were due to perturbations of "electric galvanic magnetic matter," immersed in and affected by an electric medium. Similar views, although perhaps not as all-embracing, were shared by contemporary "electricians" from William Sturgeon to Robert Were Fox.[36]

New currents continued to stimulate public interest. After observing stalactites and stalagmites in Holywell Cavern in 1807, the political radical, wealthy landowner, and amateur natural philosopher Andrew Crosse reasoned that there must be some force that countered gravity, holding mineralized water in place long enough for crystals to form. He experimented by passing a small electrical current through some water taken from the cave. After ten days, he observed small crystals of lime carbonate forming on the negative platinum lead, and he began a more systematic study of electricity in processes of crystallization and in atmospheric phenomena. Crosse's father had corresponded with Benjamin Franklin and Joseph Priestley. Crosse himself was acquainted with Coleridge, who was a friend of his brother, and he would later host visits to his home from many of the leading natural philosophers of the day, including Humphry Davy. Crosse gave one public lecture on atmospheric electricity in London in 1814. Coincidentally, Mary Shelley and Percy Bysshe Shelley also happened to attend a lecture on electricity in London in 1814, and some scholars have speculated that it was Crosse's. Some have even suggested Crosse as the model for Dr. Frankenstein. There is little evidence for the former claim, however, and the latter is even more unlikely.[37]

By 1836, Crosse had come to the attention of the British Association for the Advancement of Science (BAAS). He attended the annual meeting in Bristol and was persuaded by John Dalton to present some of his own work on crystallization, where it was well received by prominent geologists such as William Buckland and Adam Sedgwick. Geologists had recently begun to think that processes such as mineral formation, mountain building, crystallization, stratification, and cleavage involved electricity. What they found so exciting about Crosse's work was that it made geology more of an experimental science. The same year, Crosse began a series of experiments that would make him the unintentional focus of international attention and notoriety. He created a saturated solution of potassium silicate and allowed it to drip from a funnel slowly onto a piece of iron oxide that was connected to a voltaic pile. The runoff was collected and returned to the funnel, so the stone remained wet and electrified at all times. After two weeks, Crosse observed "very minute white specks or nipples" on the surface of the stone. But these were not the crystals he expected to see.[38]

Instead, Crosse was shocked to discover on the twenty-sixth day of his experiment, that tiny, perfect insects were detaching themselves from the stone and crawling around. Over the next few weeks, a hundred or so insects appeared on the stone. They were averse to sunlight. He examined them under the microscope. They had eight legs and looked like cheese mites but were larger, although he was the first to admit that he was not an entomologist. He tried another experiment, electrolyzing a solution of potassium silicate in a small glass vessel. Again, similar insects emerged from the solution after a few weeks. "Many persons have seen the insects both dead and living," Crosse concluded, "Whether the electric action has any thing to do with their birth or not I cannot say without further experimenting." The public press was not so circumspect. An unauthorized report in the *Somerset County Gazette* on December 31, 1836, described Crosse's experiments and then concluded that since a German natural philosopher had shown that rocks were comprised primarily of insect remains, "may not the germs of some of them, released from their prison-house, and placed in a position favourable to the development of vitality, have sprung to life after a sleep of thousands of years?" The article was reprinted in hundreds of newspapers and journals over the next few weeks, and caused a storm of controversy. The historian James Secord claims, "For most readers the creation of life was an experimental fact by the end of January 1837."[39]

For some, obviously, any attempt at such an action was blasphemous. One of Crosse's farmer neighbors wrote to the paper that "Andrew Crosse ought to have been hanged long ago for dealing with the devil . . . [he] raised the devil 4 or 5 times to my certain knowledge." Perhaps not. Crosse's scientific peers were unable to raise *Acarus* mites when they replicated his procedures, let alone the Prince of Darkness. John Children and Golding Bird, of the British Museum and Guy's Hospital, respectively, re-created Crosse's experimental setup and tried variations of it but failed to create living insects, a fact they presented at the BAAS meeting of 1837. Although Michael Faraday was widely reported in the popular press as having replicated Crosse's experiment, he did not. What he actually did was say that Crosse's work could not be accepted until it was further tested in a way that provided clear evidence, which Secord notes amounted "in Faraday's own terms, to a crushing dismissal." Faraday subsequently went to some trouble to counteract the effects of the fallacious report, publicly disavowing it. "With regard to Mr. Crosse's insects," he stated, "I do not think anybody believes in them here except perhaps himself and the mass of wonder-lovers." Faraday himself professed a deep Christianity—he was a member of the Sandemanians, a Protestant sect—and he actively opposed spiritualism and mesmerism. In this, he had the full agreement of Andrew Crosse, who had no use for such practices either (despite the fact that his experiments would become a staple of paranormal literature).[40]

One might not be able to create life de novo from an electric spark, but a strong enough discharge could clearly be used to stop circulation and respiration and, when the conditions were right, might be used to start them up again. Many of the same natural philosophers who explored the connections between physical, chemical, atmospheric, and animal electricity—Cavendish, Priestley, Davy, and others—also strove to understand the composition of various "airs" and the part that they played in vital processes. In 1804, Charles Henry Wilkinson suggested that respiration was a galvanic operation and that the fine structure of the lungs corresponded to the electric organs of the torpedo because both systems had the function of conveying electric charges. In his theory, positively charged air was drawn into the lungs, where the electric charge was transferred to venous blood rich in carbon. The electric charge of the blood instantly flowed to the heart, stimulating it to beat. As a consequence, the expelled air, as Volta and others had observed, had less measurable electricity. This explained "the languor and indolence we experience, when the atmosphere is but feebly charged with electricity" and "the cordial exhilaration we feel, when the barometer is high, and the air consequently charged with electricity."[41]

Electricity could thus be expected to be important in cases of "suspended animation" or apparent death, when, as Wilkinson put it, "life is not extinguished, but its influence on the animal organization merely suspended." Animals drowned in water or in hydrogen gas provided an experimental model. Although Priestley believed that hydrogen was almost as lethal as the gas of carbonic acid (carbon dioxide in our terms), Wilkinson and other researchers had tried taking breaths of hydrogen themselves without noticing ill effect. Thomas Beddoes successfully resuscitated rabbits that had been immersed in hydrogen for up to seven minutes, whereas those immersed in carbonic acid gas were "perfectly irrecoverable after one minute and a quarter." To revive victims of drowning, hanging, or the inhalation of noxious fumes, Wilkinson recommended introducing air enriched with oxygen into the lungs while sending gentle galvanic shocks through the body. Resuscitation took on wider import around midcentury, when anesthetics began to be used to render patients unconscious and insensible: "the anesthetists rapidly became, and have remained, the great resuscitators." Prior to the 1840s, "laughing gas" (nitrous oxide) and ether had been recreational drugs. Their use for surgical and dental procedures, medical historian Lloyd Stevenson suggested, probably followed from the discovery that people did not feel any pain if they were injured accidentally during an "ether frolic."[42]

In 1823, the surgeon Henry Hill Hickman wondered if he might find some way to reduce human suffering during operations and began experimenting with animals in a state of suspended animation. He did seven procedures in all, published the following year in a pamphlet entitled "A Letter on Suspended Animation." Hickman started by asphyxiating a puppy in a glass jar, surgically removed one of its ears and then allowed it to revive. The puppy soon began to breathe again, did not appear to be experiencing any pain, and healed completely within three days. Next he suffocated the same animal with carbonic acid gas and removed its other ear; once again it seemed not to suffer. For his third experiment he removed the puppy's tail. He removed a mouse's leg successfully, keeping it for a fortnight, "after which [he] gave it liberty." He exposed a dog to carbonic acid gas, rendering it apparently lifeless in twelve seconds. While occasionally inflating its lungs mechanically, he amputated the dog's leg. It did not seem to experience any pain. After seventeen minutes, he was able to resuscitate the dog, which "recovered without expressing any material uneasiness." He finished by amputating the ears of a rabbit and the ears and tail of a kitten. Although Hickman was prepared to administer galvanic fluid to restore the animals after his procedures, all revived simply by breathing open air. He was confident enough in the procedure to state he "should certainly not hesitate a moment to become the subject of it, if [he] were under the necessity of suffering any long or severe operation." In every case—at least as far as he could tell—his subjects had not suffered, and he had only experimented with animals that had already been condemned to die. Although he argued that the suffering the animals endured at his hands was less than if they had been killed by ordinary methods, his work was targeted by antivivisectionist groups.[43]

Asphyxia and anesthesia were not clearly distinguished in the mid-nineteenth-century concept of suspended animation, which took on a wide range of meanings related to, among others, drowning, asphyxia, trance, catalepsy, coma, lethargy, hibernation, hypnosis, and ecstasy. Determining the degree of pain involved in a procedure required investigation at more local scales than a whole organism in a state of suspended animation. In 1848, J. Y. Simpson, professor of midwifery at Edinburgh, reported on a series of experiments on local anaesthesia. He began by applying chloroform vapor to parts of the bodies of earthworms, showing that the anesthetized parts were unresponsive while other parts were not. The creature would simply drag around the affected part for a few minutes, gradually regaining motion and irritability in it. The medicinal leech behaved the same way, so he moved on to a small centipede, temporarily disabling some of its segments while the others continued to move around briskly. Water-newt tails and frogs' legs were immobilized for a time, although he found that he could still stimulate the latter to contraction when galvanic current was passed through them. He anesthetized the hind legs of rabbits and guinea pigs by placing them in a large airtight sac filled with chloroform vapor. "At the end of an hour the common sensibility of the limb to pinching and squeezing was much impaired; but a current of galvanism passed through it produced crying and signs of pain." Stevenson also performed experiments on himself and his pupils. In his opinion, the degree of insensibility was never enough that a patient could undergo surgery without trepidation.[44]

Although animal experiments were crucial to the development of successful methods for producing general and local anesthesia, animals were often not the immediate beneficiaries of these techniques. By signaling their pain to human observers, other animals could still be used to measure something that was otherwise inaccessible to instrumentation: a subjective state. And the fact that they could be sacrificed and autopsied with limited repercussions made it possible for experimenters to explore a continuum of antemortem, perimortem, and postmortem processes. Electricity, now divorced from animal bodies and "black-boxed" in devices like the voltaic pile, was one of the key tools in this liminal investigation. Galvanic shock was routinely used to resuscitate the near dead and helped somewhat to allay widespread fears of premature burial. Following the explorations of Nysten, the electrical excitability of muscle became one way of mapping rigor mortis and estimating the time of death for forensic purposes. The sensory or motor effects of medication could be probed with electric shock. And animals stunned or killed with electricity provided a model for lightning injuries and also, as the nineteenth century wore on, for industrial ones.[45]

FIVE

# Evolutionary Theories

## Form and Function

What role did electricity play in the emergence of inorganic form, the development of an individual organism, or the trajectory of a species? For those inclined to see them, there were meaningful patterns everywhere: geological strata resembled the layers of a voltaic pile; networks of nerves looked like Lichtenberg figures; animals of different species shared anatomical structures. The age-old human practice of disassembling animals into functional components naturally led to questions about how they had come to be assembled that way in the first place. One explanation, a perennial favorite, was that this was the handiwork of an intelligent designer. Whether one was satisfied with that account or not, there was still much more to learn.

In the 1770s, Charles Darwin's grandfather Erasmus Darwin began experimenting with the therapeutic uses of electricity and invented several electrical machines. In 1800, he built a voltaic pile within months of the arrival of Volta's news in London and began using galvanic current to treat patients. While the fluid that animated living fibers must be similar to animal electricity, Erasmus Darwin did not believe that they were one and the same. In *Zoonomia* (1794), he argued that a "living spirit, or spirit of animation . . . resides throughout the body, without being cognizable to our senses, except by its effects."

> The similarity of the texture of the brain to that of the pancreas, and some other glands of the body, has induced the inquirers into this subject to believe, that a fluid, perhaps much more subtile than the electric aura, is separated from the blood by that organ for the purposes of motion and sensation. When we recollect, that the electric fluid itself is actually accumulated and given out voluntarily by the torpedo and the gymnotus electricus, that an electric shock will frequently stimulate into motion a paralytic limb, and lastly that it needs no perceptible tubes to convey it, this opinion seems not without probability; and the singular figure of the brain and nervous system seems well adapted to distribute it over every part of the body.

Erasmus Darwin also developed an associationist psychology from the work of Hartley and then applied it to ontogeny and phylogeny. Reproduction from a single living filament gave rise to a new animal that was similar to its parent in form and inclination.

> From this account of reproduction it appears, that all animals have a similar origin, viz. from a single living filament; and that the difference of their forms and qualities has arisen only from the different irritabilities and sensibilities, or voluntarities, or associabilities, of this original living filament; and perhaps in some degree from the different forms of the particles of the fluids, by which it has been at first stimulated into activity. And that from hence, as Linnaeus has conjectured in respect to the vegetable world, it is not impossible, but the great variety of species of animals, which now tenant the earth, may have had their origin from the mixture of a few natural orders.[1]

Étienne Geoffroy Saint-Hillaire, Erasmus Darwin's much younger contemporary, also wished to explain the anatomical similarities to be found between animals of different species. Physician turned zoologist, Geoffroy went to Egypt in 1798 as part of the scientific expedition that accompanied Napoleon's invading force. There he made a practice of drawing, dissecting, and then preserving any vertebrates he could obtain from the locals; eventually these specimens returned with him to the National Museum of Natural History in Paris. Geoffroy's chief concern was to explain the anatomical curiosities that he found in fish: the mechanism by which the puffer *Tetraodon* became inflated, the appendix of the shark and ray, organs in the catfish that appeared similar to human lungs, the peculiarities of the elephantfish.

During the siege of Alexandria in August 1801, Geoffroy had the opportunity to compare a torpedo with an electric catfish from the Nile, and he became so immersed in the "problems of natural philosophy" for a few weeks that he lost track of "the bombings, the local fires, the surprises of the besiegers, and the plaintive cries of the victims." Musing over the anatomies of the two strongly electric fish, Geoffroy imagined an elaborate interplay of forces and subtle fluids. Caloric became divorced from matter in the lungs and circulated through the nervous system as nervous fluid. Muscular contraction was caused by an arterial stream of caloric encountering a stream from the nerves. Two molecules of caloric could join through an electric process, losing their "expansibility" and becoming a molecule of carbon, and combustion was the inverse process, also electric, by which carbon could return to caloric. Generation was caused by the union of male and female streams of caloric. In a letter to Georges Cuvier, Geoffroy wrote, "One can explain all galvanic, electric, and magnetic phenomena, the nervous fluid, germination, development, nutrition, generation . . . I say also that one can explain the intellectual function by physics, something that I have been told has already been done, and that I will dare to undertake."[2]

Georges Cuvier, Geoffroy's colleague at the museum, was also a comparative anatomist. More pragmatic and politically astute than Geoffroy, he had made the decision to retain his post in France rather than adventuring abroad with Napoleon's forces. Cuvier's primary concern was to establish the functions for which anatomical structures were adapted, and he maintained that animal life was divided into four "embranchements": vertebrates, mollusks, articulates, and radiates. He used the arrangement of a creature's nervous system to determine in which embranchement it belonged. Whatever structural similarities that might be found within or across these divisions were due merely to the need to serve a similar purpose. There was no space in Cuvier's scheme for transitional forms between the categories. Animals were assumed to have been given whatever properties they required for their particular niche by a benevolent Creator.

The question of function was also central to the natural theological tradition, especially in Britain. In his *Natural Theology* (1802), William Paley described a multitude of mechanisms that allowed animals to attack one another and defend themselves, ranging from horns, teeth, beaks, talons, stings, and prickles to "(the most singular expedient for the same purpose) the power of giving the electric shock." Because any of these devices could serve as well as any other, Paley concluded that "*variety* itself, distinct from every other reason, was a motive in the mind of the Creator, or with the agents of his will." In the early decades of the nineteenth century, natural theologians would wholeheartedly embrace Cuvier's arguments, using them to support claims that any adaptation that an organism exhibited to a niche was evidence of intelligent design.[3]

The electric discharge of strongly electric fish was almost always given a functional explanation, although the details occasionally varied. Charles Henry Wilkinson, for example, argued in 1804 that the slow-moving electric eel required the ability to produce luminous sparks in order to attract and delay faster-moving prey. Geoffroy was unusual in downplaying the functional aspects of electric organ discharge to focus on morphological ones instead. His comparative anatomical investigations of the torpedo and other rays suggested to him that rays that were not capable of producing electrical discharges nevertheless possessed organs analogous to the electric ones in the torpedo. Back in France, he was able to make a close comparison of his Egyptian specimens with those Cuvier had assembled. In 1807 Geoffroy concluded that—contrary to the claims of Cuvier—the organs of fish were deeply analogous to those of other vertebrates. Instead of giving fish bones specialized names, as Cuvier did, Geoffroy named them according to equivalent bones in mammals. If bones in a particular species were sometimes rudimentary, modified in some way, or fused together, Geoffroy was willing to overlook the fine details in favor of finding an overall generalization. The connections between bones were invariant across species, and for him that was evidence of an underlying blueprint or plan. In twentieth-century terms, Geoffroy was a "lumper" and Cuvier a "splitter."[4]

Geoffroy's perspective was close to the transcendental anatomy espoused by his German contemporaries like Johann Wolfgang von

Goethe, Lorenz Oken, and Johann Friedrich Meckel. For them, too, animals showed a unity of plan and could be arranged on a continuous linear scale. The German biologists also argued that individual development paralleled stages on the chain of being. Oken, for example, claimed that a developing animal would pass through each prior stage of the animal kingdom and that adult animals were merely "the persistent foetal stages or conditions of Man." Where Oken saw this developmental progression in terms of the cumulative addition of organs, Meckel instead formulated it with regard to coordination and specialization: a simple animal consisted of poorly integrated components that were much alike, whereas more advanced animals had distinct organs that were specialized and worked in concert with one another. Meckel believed that the organs themselves contained an innate capacity to be modified, a process he imagined in electrochemical terms. "The earliest development of the embryo must occur at the edges of the primitive spinal column analogously to the phenomenon of electricity and can perhaps be conceived best as follows: The two plates which extend from each side of the spinal column, similarly to the leaves of the Diana tree of the surface of a copper plate, must on account of the lateral duality in terms of which they are constructed, have opposite polarity on the surface and on each of the ends; and on account of their polarity the two opposite plates have a tendency to approach one another and unite." (Under the right conditions, a metallic salt solution can be reduced to metal crystals with a treelike form. The silver version was known to alchemists as "Diana's tree." Around the turn of the nineteenth century, William Cruickshank and Johann Wilhelm Ritter had independently discovered that galvanic electricity could speed up the process of forming such dendritic structures, and it was this that Meckel had in mind when he imagined embryonic development.)[5]

Karl Ernst von Baer took up where Meckel left off, rejecting the former's recapitulationism while maintaining a similar developmental mechanics. Trained in the physics of electricity and magnetism, as well as comparative anatomy, von Baer also imagined that electrochemical processes played a vital role in embryonic development. The historian Timothy Lenoir interprets von Baer as claiming that the successive alignment of heterogeneous materials along one axis of the egg created a galvanic current, and thus a magnetic field encompassing it. This magnetic field, in turn, directed the further growth of the embryo. Whether or not one believed in recapitulation—the idea that the developing embryo took on the forms of ancestral organisms—the German biologists and Geoffroy and his followers were in agreement about the existence of an unbroken, related series of animals and of common organ components. Geoffroy's *Philosophie anatomique* (1818–22) argued that people, apes, lizards, mice, and fish shared a "unity of composition" and were built from the same stock of "organic materials." This shared belief made it much easier to imagine a historical, and ultimately an evolutionary, progression.[6]

## Electric Agency

The anonymous publication of *Vestiges of the Natural History of Creation* in 1844 brought evolutionary debate to the mainstream. It argued that the development of the universe, the origin and evolution of life and its precursors, and the ultimate perfection of humankind were all inevitable, foreordained by a natural law that was laid down by a hands-off God. Having established the way that the universe worked, He didn't need to complicate things with "fiats," "special miracles," "interferences," or other kinds of supernatural action. *Vestiges* took a story that had previously been the province of atheists, deists, working-class socialists, and French revolutionaries, and recast it in the polite and devout language of the Victorian middle class. The book became hugely popular and controversial, selling more than 23,000 copies over the next decade and a half and reaching an audience estimated at more than 100,000 readers.

The argument in *Vestiges* drew on recent findings from a range of natural and human sciences—anthropology, astronomy, biology, chemistry, geology, phrenology, physics, and political economy—which made it all the more difficult for readers to try to guess who the author was, although that soon became a popular parlor game. A case could be made for practically anyone who specialized in one of the subjects the book covered, or for any of the authors it cited: Charles Lyell, Charles Darwin, George Coombe, William Carpenter, Charles Babbage, Richard Owen. "When controversy was at its height," historian James A. Secord writes, "the best-informed readers thought *Vestiges* might be by Lord Byron's daughter, the Countess Ada Lovelace. Lovelace was known for her friendship with Babbage (whose calculating engine featured so prominently in the book); her acquaintance with Crosse (discoverer of the electrical creation of life); and her ambitions to revolutionize natural philosophy."

What a cast of characters! And what an exercise in counterfactual history: to imagine if Ada Lovelace really had written *Vestiges*. In retrospect we've mostly assigned these people to their own embranchements: making them protagonists in the history of mathematics or medicine, political history or women's history. Needless to say, the historical actors didn't see themselves that way. The issues that *Vestiges* raised sat squarely in domains of contested authority, between religion and natural philosophy, experts and amateurs, men and women, and the upper, middle and working classes, embroiling practically everyone in Victorian society in the argument. The actual author, although he never admitted it publicly, was a middle-class publisher named Robert Chambers. In the 1830s, he joined his brother William to produce a successful weekly, *Chambers's Edinburgh Journal*, filling each issue with science, popular literature, and curiosities. In the process, Chambers received a wide and eclectic education, using the pages of the *Journal* to work through and develop his own ideas about the material he was covering. Much of this would later appear in *Vestiges* and its sequel, *Explanations*, which was published the following year. Secord argues, "The first step in reading *Vestiges* and *Explanations* . . . is to forget about Robert Chambers, for only then can we recapture the sense of dislocation experienced by the original readers."[7]

The impetus for *Vestiges* was astronomer John Pringle Nichol's 1837 *Views of the Architecture of the Heavens*, which related the nebular hypothesis in popular terms. On this view, the force of gravity led to the aggregation of matter around stellar nuclei, resulting in the detachment of more or less solid stars from dense clouds of gas. Chambers was impressed that the universe's physical origins could be described in such a lawlike and progressive manner, and *Vestiges* was his attempt to show that natural law could account for organic nature, too. The "Almighty Deviser" worked algorithmically, to use the word in a sense that became current toward the end of the century. Quoting an extended passage from Charles Babbage, *Vestiges* described how it was possible for a computing engine to be programmed to arbitrarily change its mode of operation at a predetermined time. Chambers used this observation to support a somewhat inconsistent, recapitulationist phylogeny. "Unquestionably, what we ordinarily see of nature is calculated to impress a conviction that each species invariably produces its like." Such a view, however, was due to the limited perspective of humankind. Both the ages before history and those yet to come were unknown.[8]

In *Vestiges*, electricity was a key force in the emergence of inorganic form, the creation of the building blocks of life, organic development, the structure of the nervous system, and processes of thought. Crystals of frost on a window were patently inorganic, yet their forms clearly resembled those of plants, something that was apparent to the "simplest rustic observer." The Diana Tree, although made of silver, perfectly resembled a shrub, and its growth could be manipulated by electricity and magnetic fields. Lichtenberg figures created with positive electricity resembled trees, those created with negative electricity looked like bulbous or spreading roots. This seemed like clear evidence that the form of plants was somehow related to electric energy. It was also noteworthy in this regard that the lower atmosphere is positively charged with respect to the earth. Negatively charged water was found to inhibit plant germination, whereas positively charged water promoted it. And so on. In *Explanations*, Chambers quoted *Leithead's Electricity* (1837) in support of the claim that the passage of electricity through a body stops when the force of the electric fluid is exactly counterbalanced by the resistance of the material. At that point, the electric fluid seeks a different path, and it is this characteristic that was thought by both William Leithead and Chambers to give plants their various regular forms, and perhaps even to operate on animal embryos.[9]

The emergence of organic from inorganic forms also required electricity. In *Vestiges*, the most basic kind of organic being was a "globule" that contained other globules. Since a French physiologist was credited with creating globules by subjecting albumen to electrical current, the question of "the commencement of animated creation by the ordinary laws of nature" seemed to devolve to the problem of creating albumen inorganically. That looked like a straightforward problem of chemical synthesis. The electrical formation of something sort of cell-like was one thing, but *Vestiges* had even stronger proof that the organic could emerge from inorganic precursors in the presence of electricity: Andrew Crosse's generation of *Acarus* mites. Chambers also cited and carried out a correspondence with W. H. Weekes, who had reported a replication of the Crosse experiments to the London Electrical Society in March 1842. Weekes's insects were observed to mate with one another, to feed in the fluid, and occasionally to eat their own kind. Reproduction, nutrition, and conflict were hallmarks of the living, and the emergence of the "*Acarus Crossii*" might simply be the final stage in a long chain of transformations of the solution caused by the electrical current. That *Vestiges* endorsed the work of Crosse and Weekes was taken as evidence in favor of the author being someone like Harriet Martineau or Ada Lovelace, both of whom had publicly embraced the electrical creation of life.[10]

As a universal force, electricity must also be concerned with mental action, and *Vestiges* concluded that "life and mental action must everywhere be of one general character." The volume or mass of a creature's nervous system indicated how powerful its mental abilities were. In mollusks and crustaceans it seemed to be little more than a ganglion running the length of the body and sending out lateral offshoots. Vertebrates had a brain and spinal cord with a branching network of nerves. The lowly ray fared poorly in *Vestiges*, having only "the first faint representation of a brain in certain scanty and medullary masses, which appear as merely composed of enlarged origins of the nerves." Electricity—which could also take the form of magnetism, heat, or light—could stimulate the nerves of a dead body to action or cause the brainless remains of a recently killed animal carcass to restart the processes of digestion. This showed that brains were little more than galvanic batteries. "Nor is this a very startling idea, when we reflect that electricity is almost as metaphysical as ever mind was supposed to be. It is a thing perfectly intangible, weightless." In a footnote, Chambers drew one further inference: if thought was electric, then the speed of thought must be equal to the speed of electricity (and light), conducted at 192,000 miles per second. This was more than fast enough that ordinary muscular movements would appear to coincide with mental plans to move.[11]

James Secord argues that our reading of *Vestiges* has been distorted by treating it as a failed attempt to provide an evolutionary account of the origin of species, a work of popular science superseded by Darwin's own *On the Origin of Species* in 1859. What Robert Chambers actually intended to accomplish was much broader: nothing less than "a vision of nature appropriate to the industrial age and the middle classes," one that covered everything from the origins of stars to the emergence of human mental life. By reforming Victorian civilization, Chambers hoped to make it susceptible to the same progressive forces that he believed governed everything else. Secord traces our own fascination with big histories—reflected in the success of popular works like Stephen Hawking's *Brief History of Time* and Steven Pinker's *Language Instinct*—to the reception that *Vestiges* received in the mid-nineteenth century, rather than to any pride of place enjoyed by cosmologists or Darwinian scientists in our own day. "*Vestiges* created a space for debate about natural law, setting the stage for the controversy over Darwin's *Origin*—a book that in many ways presented its arguments as a response to what Chambers had done."[12]

## Transmutation of Species

Electricity did not play nearly as central a role in Darwin's evolutionary thinking as it did in *Vestiges*, but strongly electric fish presented him with a puzzle that he revisited throughout his life. Darwin's experiences with electricity were typical of his day and age. As a schoolboy, he and his elder brother set up a chemical laboratory in an old scullery and tried their hand at analyzing minerals, metals, and crystals, although they couldn't afford the equipment that they needed to replicate Davy's electrolysis of water. While a student at Edinburgh in 1826, Darwin enjoyed T. C. Hope's lectures on electricity and was glad that he remained long enough to hear them before heading home at the end of term. A few years later, he read John F. W. Herschel's *Preliminary Discourse on the Study of Natural Philosophy* (1830), which described the action of physical forces including electricity and magnetism and emphasized the induction of natural laws from observations. He read Humboldt's *Personal Narrative* "over and over again" and developed an ardent desire to travel as Humboldt had. When the opportunity came to voyage on the *Beagle*, Darwin leapt at it.[13]

Preparing for the *Beagle* voyage, Darwin attended a lecture on lightning conductors at the Athenaeum given by a Mr. Harris. The ship was to have copper plates in the masts and yards that were connected to the sea. Harris demonstrated their function with a model consisting of an electrical machine, metal thundercloud, toy boat, and a tub of water. "We shall be the means of trying & I hope proving the utility of its effects," Darwin wrote in his diary. The following summer, off the southern coast of Uruguay, he watched as "the darkness of the sky [was] interrupted by the most vivid lightning.—St. Elmo's fire. The tops of our masts & higher yards ends shone with the Electric fluid playing about them." At St. Jago in the Cape Verdes in January 1832, Darwin had a several opportunities to observe small cuttlefish trapped in tide pools, although they seemed quite smart and were not easily caught. He was particularly interested in their ability to change color, apparently at will. He described the hues that passed rapidly over their bodies: "brownish purple," "yellowish green," "French gray," "bright yellow," "hyacinth red," "chestnut brown." "Any part, being subjected to a slight shock of galvanism," he noted, "became almost black." At Montevideo, Uruguay he toured a house that had been recently struck by lightning, noting that the bell wires had melted and fused, and "In one of the walls the electric fluid exploded like gunpowder, & shot fragments of bricks with such force as to dent the wall on the opposite side."[14]

In November 1837, while he was privately working on his theory of evolution, Darwin's future sister-in-law Sarah Elizabeth Wedgwood wrote to him to ask if he was going to see "Mr Crosse's animals" for himself, as a "cargo" of them had arrived in London. Darwin had little use, however, for the mesmerists and electricians who multiplied connections willy-nilly between electricity, magnetism, and vital and neural forces. Writing to his second cousin William Darwin Fox in December 1844, he ridiculed Harriet Martineau's mesmeric cures (although he actually liked her as a person and found her interesting). He sarcastically suggested that the "experimentum crucis" would be to see if a mesmerist could put some cats into a stupor because they were reputed to be "so electrical." Darwin had returned from his round-the-world voyage with a debilitating illness—post hoc diagnoses include a wide range of possibilities such as Chagas' disease, hypoglycemia, multiple allergy, hyperventilation, and neurasthenia—and he briefly sought relief in what he described to Joseph Dalton Hooker as "a great piece of quackery, viz twice a day passing a galvanic stream through my insides from a small-plate battery for a half an hour." The treatment apparently didn't work for him. Darwin didn't mention galvanic therapy after 1846, although he remained an invalid for the rest of his life.[15]

Darwin finished a version of his manuscript in the summer of 1844. He wasn't quite sure what to do with it, however, as it flew in the face of orthodox Victorian natural history. He tried broaching the topic with a few close friends and found them to be receptive. "Having kept his secret for so long," Janet Browne writes, "Darwin naturally felt he was the only man in Britain who possessed a fully worked out theory of transmutation. That same autumn he was proved wrong—devastatingly wrong." *Vestiges* was published in October. The surgeon and botanist Joseph Dalton Hooker read it and wrote Darwin to tell him that he was "delighted" with the book, although he could not agree with the conclusions and found errors aplenty. By the time that he received Hooker's letter, Darwin had already read the book, too, in the library of the British Museum. Not surprisingly, Darwin found himself "somewhat less amused" with *Vestiges*, concentrating on what the author had to say about the transmutation of species and more or less ignoring much of the rest. He treated it, Secord writes, "not as a sweeping cosmological narrative but as a botched version of his own manuscript." When authorship of *Vestiges* was attributed to him, Darwin felt he "ought to be much flattered & unflattered." Learning from the reception of *Vestiges*, Darwin put off publication of his own work on species, immersing himself in an eight-year-long study of barnacles.[16]

## Puzzles for Darwin

Darwin's unpublished essay of 1844 asked "whether any particular corporeal organs, or the entire structure of any animals, are so wonderful as to justify the rejection *primâ facie* of our theory." The eye, for example, seemed too complicated to "have been acquired by gradual selection of slight, but in each case, useful deviations," yet Darwin thought that a case could be made. He also discussed the intermediate forms that may have been taken by bats, the development of the swim bladder in fish from part of the ear, and a few other apparent difficulties.

One potential objection that does not seem to have occurred to him for some time was the ability of strongly electric fish to shock. He had known about such fish for years. They were discussed by Humboldt (vividly and at length), Paley, Lyell, J. F. W. Herschel, Richard Owen, William

Benjamin Carpenter, and Alfred Russel Wallace, among many other authors whom Darwin had read. A forty-inch-long, one-eyed, live electric eel was to be seen at the Adelaide Gallery in London from 1838 to 1842, where it was a star attraction. By the end of 1856 at the latest, however, strongly electric fish were starting to trouble Darwin. In a letter to Huxley he wrote, "Cases of organs in which there is no apparent passage or transition from other organ: or still better, if such transition can be shown in an unexpected manner. E.G. Electrical organs in Fish, seem to be really new organ & not any other changed . . . I require passages, but I always give all the facts which I can collect, hostile to my notions."[17]

In the 1857 notes that would eventually become chapter 6 of *Origin*, Darwin cited Owen's Hunterian Lecture on the nervous system of fishes, observing that electric organs created a "special difficulty" for him. The research on strongly electric fish reviewed in Carpenter's 1854 *Principles of Comparative Physiology* suggested that many questions about the origin and use of electric organs remained to be answered—particularly as James Stark and John Goodsir had each shown that rays, which apparently could not discharge electric shocks, nevertheless had organs that were anatomically similar to those in the strongly electric fish. "We are at present too ignorant to speculate on the stages by which these organs, now affording such a powerful means of defence to the Torpedo & Gymnotus, may have been acquired," Darwin wrote. Worse, electric fish belonged to widely separated subclasses and thus must have diverged from one another "at an immensely remote period & therefore can hardly owe this similar anomalous organ to community of descent."[18]

Darwin did not manage to resolve the problem before *Origin* was published in 1859. In the first edition he acknowledged that "it is impossible to conceive by what steps these wondrous organs have been produced." Electric organs only appeared in a few, widely separated species. If extant strongly electric fish had inherited the organ from "one ancient progenitor," they should be closely related to one another, but they clearly were not. The geological evidence weighed against the hypothesis that electric organs had once been widespread among fish and subsequently lost in their descendants. "I am inclined to believe," Darwin wrote, "that in nearly the same way as two men have sometimes independently hit on the very same invention, so natural selection, working for the good of each being and taking advantage of analogous variations, has sometimes modified in very nearly the same manner two parts in two organic beings, which owe but little of their structure in common to inheritance from the same ancestor." One should not imagine abrupt change simply because it was difficult to imagine what steps might have been taken: nature does not make leaps. "Or, as Milne Edwards has well expressed it, nature is prodigal in variety, but niggard in innovation. Why, on the theory of Creation, should this be so? Why should all the parts and organs of many independent beings, each supposed to have been separately created for its proper place in nature, be so invariably linked together by graduated steps? Why should not Nature have taken a leap from structure to structure? On the theory of natural selection, we can clearly understand why she should not; for natural selection can act only by taking advantage of slight successive variations; she can never take a leap, but must advance by the shortest and slowest steps." If the theory of natural selection were not to be rejected out of hand, it would be necessary to find a series of short steps that led from non-electric ancestors to strongly electric eels, rays, and catfish.[19]

The first edition sold out on the day that it was published, and a second edition was printed in January 1860. Many reviewers had difficulty accepting Darwin's (lack of an) account of the electric organs in fish. In a letter to Darwin, the physician Henry Holland said that he could not imagine even an approximate solution to the problem posed by electric animals. Referring to Darwin's description of electric organs as similar to the independent inventions of two men, naturalist William Jardine wrote in an anonymous review, "We do not think Mr Darwin intends to lower the power or attributes of Deity . . . [but] to say the least, it is not the language in which we have been accustomed to see these subjects treated." Bishop Samuel Wilberforce didn't pull any punches on the question of electric organs. He could see no possible solution and upbraided Darwin for dealing with them "in a mode most unsatisfactory by one promulging a new theory of creation." Darwin's plea of ignorance was "a solution which could of course equally make the scheme it is intended to serve compatible with any other contradiction." In fact, piscine electric organs bedeviled Darwin through successive editions of his masterpiece and continued to provide his critics with ammunition long after his death.[20]

At the end of 1860, Darwin received a letter from Robert M'Donnell, an Irish surgeon and naturalist. He couldn't place him; if he was the bearded fellow whom Darwin saw at the British Association meetings, then Darwin feared he was "rash & wild." Nevertheless, he brought some good news. M'Donnell was inspired to begin dissecting common rays by reading about electric organs in *Origin*. He argued that the organ in the ray that Geoffroy and others had identified as homologous with the electric organ of the torpedo could not be, because the same organ was also present in the torpedo, where it had nothing to do with the production of electricity. The organ that Stark had identified in the tail of the skate and other rays could not be the homologue, either, since the torpedo's electrical organ was in its head rather than its tail. But M'Donnell had identified a different organ in the head of the non-electric ray that was a much better candidate for homology with the torpedo's electric organ.

In 1861, M'Donnell published a paper in the *Natural History Review* that spelled out the new implications of the study of homologies for those who had not prejudged the theory of natural selection. A complex organ that served a special function and was apparently absent in closely related organisms (such as the electric organs of fish) was problematic for the theory, as Darwin himself had argued. In contrast, "The presence of modified, atrophied, or rudimentary organs, constitutes one of the strongest arguments in favour of Mr. Darwin's theory." M'Donnell suggested that the non-electric organs found in both the head and tail of the common ray had homologues in strongly electric fish. The tail organ, which was already well known from the work of Stark, Goodsir, and others, was homologous to the electric organ of the electric eel (which was in its tail). The head organ that M'Donnell had discovered was homologous to the electric organ of the torpedo. Although M'Donnell was not yet convinced that Darwin was right, he had lessened the force of one of the crucial objections to natural selection.[21]

Darwin was pleased. Writing to Lyell, he described how M'Donnell of Dublin—"first rate man"—had gone looking for transitional forms that may have eventually resulted in the electric organs, "& is it not satisfactory that my hypothetical notions shd have led to pretty discovery"? It was a fine piece of evidence to rebut skeptics like Wilberforce or Owen, and Darwin mentioned M'Donnell's work in his letters to other correspondents around that time. The third edition of *Origin* was published in April 1861 without substantial change to the section on electric organs, but the section was revised and expanded in the fourth edition

(1866) to include M'Donnell's findings, among other things. The difficulty of electric organs remained, "but when we look closer to the subject," Darwin wrote, "we find in the several fishes provided with electric organs that these are situated in different parts of the body,—that they differ in construction, as in the arrangement of the plates, and . . . in the process or means by which the electricity is excited,—and lastly, in the requisite nervous power (and this is perhaps the most important of all the differences) being supplied through different nerves from widely different sources. Hence in the several remotely allied fishes furnished with electric organs, these cannot be considered as homologous, but only as analogous in function." Thus the problem had become the difficulty of finding the steps by which electric organs evolved in each of the separate fish that possessed them. Darwin argued that it was easier to believe in descent with modification and selection, however, than that "organic beings have been formed in many ways for the sake of mere variety, like toys in a shop."[22]

Darwin struggled with "designed laws" and "undesigned results" over the course of the 1860s. At the beginning of the decade, he couldn't believe that God designed the death of a good man struck by lightning, or, for that matter, that He intervened when a gnat was eaten by a swallow. "If the death of neither man or gnat are designed, I see no good reason to believe that their first birth or production shd. be necessarily designed," he wrote to Asa Gray, "Yet . . . I cannot persuade myself that electricity acts, that the tree grows, that man aspires to loftiest conceptions all from blind, brute force." In 1863, he confessed to Hooker, "I have long regretted that I truckled to public opinion, and used the Penteteuchal term of creation, by which I really meant 'appeared' by some wholly unknown process. It is mere rubbish, thinking at present of the origin of life; one might as well think of the origin of matter."

While Darwin and many of his colleagues found the theory of natural selection to be more convincing and better supported year by year, for some there could never be enough evidence to override the belief in an intelligent designer. It was possible to be both an evolutionist and theist, as were Asa Gray and Charles Kingsley. Darwin himself inclined more to agnosticism, a term his friend Huxley coined in 1870 and one Darwin, albeit retroactively, thought well suited to his own frame of mind. The erosion of faith in the nineteenth century was not caused by evolutionary theory or by science more generally, or limited to evolutionists or to "scientists" (the word was invented in the 1830s and began its steady rise to prominence after 1860) but was part of a wider secularization of Britain and Europe.[23]

The powerful natural laws described in *Vestiges* and *Origin* left little scope for divine activity and were increasingly used by nonbelievers to support arguments against the existence of God. One response was to argue that natural laws were simply God's way of acting in the world. This was the tack George Campbell, geologist and supporter of the Church of Scotland, took in his 1867 *Reign of Law*. When we consider the steam engine, the telegraph, or Babbage's calculating machine, we can see that they are "simply contrivances for bringing natural Forces into operation." Campbell claimed that the "machinery of Nature" was no different, save "the ultimate agency is concealed from sight." He used the electric organs of fish to provide an extended example. Their exact mechanism was still being worked out, but their purpose was clear: offense and defense. "Creatures which grovel at the bottom of the sea or in the slime of rivers, have been gifted with the astonishing faculty of wielding at their will the most subtle of all the powers of Nature." But electric organs obeyed the same laws as the electric telegraph. Everything in nature was governed by law. If the results seemed extraordinary, Campbell argued, it was because "some common law is yoked to extraordinary conditions, and its action is intensified by some special machinery."[24]

Other critics sought to make a space for the emergence of complex structures that were not designed. The philosopher George Henry Lewes suggested that it "requires a strong faith to assign Natural Selection as the cause" of phosphorescent organs in insects and electric organs in fish. Why not assume instead that these capabilities could spontaneously evolve in organisms facing similar conditions? Once present, natural selection could account for their persistence if they were of some benefit to the animal. Darwin was perplexed. "I could almost as soon admit that the whole structure of, for instance, a woodpecker, had thus originated," he wrote to Lewes, "that there should be so close a relation between structure and external circumstances which cannot directly affect the structure seems to me to [be] inadmissible." But Darwin admitted that "I have to make, in my own mind, the violent assumption that some ancient fish was slightly electrical without having any special organs for the purpose." He further speculated that "the so-called electric organs, whilst in a condition not highly developed, may have subserved some distinct function."

Writing to Alfred Russel Wallace around the end of 1868, Darwin sent him Lewes's paper. Wallace responded by saying that Lewes had missed the import of the electric organs. Lewes recognized that their position and form differed widely from one fish to the next, although their fine structure was almost identical. But this was because electric organs arose from the modification of muscle tissue. "If electrical and luminous organs were always identical in form and position as well as in structure, it would be a powerful argument in his favour," Wallace wrote, "but as it is, I do not see that it proves anything but that the required special variation of an (almost) identical tissue occurs very rarely, and has still more rarely occurred at a time and under conditions which rendered its accumulation useful to the animal, in which case alone it would be selected and specialized so as to form a perfect electric or luminous organ."[25]

Late in life, Darwin speculated that the struggle of some fishes to free themselves from parasites may have generated a mild electricity (via muscular activity) that annoyed the freeloading organisms and caused them to detach themselves. In a parable appended to a letter to George John Romanes, he speculated that the offspring of the struggling fish "gradually profited in a higher degree and in various ways by discharging more electricity and by not struggling." The puzzle of electric organs would not be solved in the nineteenth century, however, and it continued to serve as fodder for Darwin's opponents, such as the Reverend Francis Orpen Morris, who published several anti-Darwinian books in the late nineteenth century.[26]

To this day, critics of Darwinian natural selection take a range of positions on the electric organs of fish that Darwin would have found quite familiar, especially since he was the originator of many of the objections. The teacher's guide for a biology laboratory text published by the Christian Liberty Press (2005) suggests that students should search in the library and laboratory for biological structures and functions that "submit readily to the creationist rationale." One example is the electric organ in fish. "The creationist reflects, 'Wouldn't it be amazing if an electric shocking organ arose by chance just once! Are we to believe that it happened by natural selection in repeated instances?'" An article in *Creation* magazine suggests that the convergent or parallel evolution of electric fish requires "mathematical improbabilities" that are too high to accept. Brian Thomas, a science writer at the Institute for Creation Research ("Biblical, Accurate, Certain") writes that "it takes a great—and unfounded—faith in nature to maintain that . . . subtle variations [in weakly electric fish] have anything to do with ultimate origins, or that remarkably specified sensory organs emerged 'naturally' at multiple times and in different organisms." Many non-Christian creationists draw on similar arguments against Darwinian natural selection. The Muslim creationist works of Harun Yahya (a pen name for Adnan Oktar), for example, argue that the piscine electrical system is too complex to have evolved in a piecemeal fashion, and a system that was not fully functional would not have conferred any benefit on its possessor.[27]

## A Small Mass of Matter

Strongly electric fish continued to pose puzzles even for those who were not concerned with the transmutation of species. Humphry Davy's interest in fish, for example, continued to the end of his life. In 1828, his book *Salmonia* appeared anonymously, a collection of nine dialogues about fishing, set beside the streams and lakes where Davy himself had fished with his friends, his brother John Davy, and his apprentice Michael Faraday (who went on the trips but spent his time studying plants and rocks instead of fishing). By the autumn of that year, Davy was traveling in Italy. With the help of the British consul at Trieste, he obtained a pair of live torpedoes. He passed their shocks through a silver spoon into an electrometer, but they did not budge the needle of the instrument. Wetting his hands in salt water, he put himself into the circuit to confirm that current was flowing through it, and felt the shock to both elbows. He hypothesized that the animal's shock was instantaneous, whereas the instrument required the current to continue for a time before it would register. A weakly charged Leyden jar did not register on his electrometer, either, whereas a small voltaic battery did. Davy wished that he had a much more powerful electric eel to test, suspecting that animal electricity would prove to be "of a distinctive and peculiar kind" when compared with common or voltaic electricity or magnetism. He was ill, however, and the experiments would be his last. He exhorted his brother to continue the work.[28]

John Davy joined Humphry in Malta in March 1829. Although Humphry was too sick to study the torpedoes himself, John "amused him, day after day, with the results of [his] dissections" until Humphry's condition worsened. Humphry Davy died at the end of May, and John did not have an opportunity to continue the research for a few years. He started with an experiment that Humphry had recommended before his death. In 1820, Humphry Davy, Thomas Johann Seebeck, and Dominique François Jean Arago had independently discovered that an electric current attracted iron or steel filings and needles. The same year, Arago also reported that a steel needle could be permanently magnetized if it were placed along the axis of a current-carrying coil of wire. Iron needles could be temporarily magnetized under the same conditions. Humphry Davy had wanted to test whether the shock from an electric fish could also magnetize a needle if it were passed through a coil. John confirmed that it could. He then showed that the torpedo's shock could also deflect the needle of a galvanometer, a device which measured electrical current. He confirmed his brother's (and John Walsh's) observation that it would not register on a common electrometer, however, and he could not get the spark to jump a gap, no matter how minute. John Davy also successfully used the fish as a source of current for other electrochemical experiments. After detailed anatomical work, he concluded that "the question remains unanswered, What is the cause or source of the electricity? Here analogy fails entirely; none of the ordinary modes of excitement appear to be at all concerned; neither friction, nor chemical action, nor change of temperature, nor change of form . . . The smallest [torpedo] which I have employed in my experiments weight only 410 grains . . . Yet this small mass of matter gave sharp shocks, converted needles into magnets, affected distinctly the multiplier [i.e., galvanometer], and acted as a chemical agent, effecting the composition of water, &c. A priori, how inconceivable that these effects could be so produced!"[29]

John Davy was not the only person studying torpedoes in the 1830s. In 1831, Swiss physicist Daniel Colladon used a galvanometer to measure the direction and strength of electric currents in torpedoes. There was still some question about whether animal electricity was identical to electricity derived from other sources. Faraday thought it was. In 1833, he hypothesized that the torpedo would generate heat while discharging an electric shock, and that this could be measured with a thermo-electrometer. John Davy did so the following year, and three years later, Santi Linari built a circuit with a live torpedo, a bismuth and antimony thermo-electric element, and a sensitive air thermometer. He, too, showed that the electric currents generated by the torpedo did indeed have "calorific properties." Linari also demonstrated that the torpedo was capable of generating a perceptible spark—something John Walsh had shown for the electric eel in the 1770s, but which had once again become a matter of some debate. Torpedoes, however, did not survive for long in captivity, so those who wished to experiment with them on a regular basis had to live near a place where they were plentiful and caught daily for food, like Naples.[30]

## An Inquiry of Surpassing Interest

Researchers in London lacked ready access to live torpedoes for experimentation, but at the beginning of September 1838, Michael Faraday noted that a live electric eel had arrived the previous month. Three years earlier, Faraday had petitioned the Colonial Office for their assistance in procuring *Gymnotus* specimens with the encouragement of Humboldt. Humboldt had also provided him with detailed instructions about where the fish could be captured (small rivers flowing into the Orinoco), when to transport them (from Surinam in the summer), what they ate (small fish, or meat that was cooked but not salted), and how to ship them (in a net-covered, wood trough, with fresh water changed every three or four days). "It is, however, *important* that the animal should not be tormented or fatigued, for it becomes exhausted by frequent electric explosions." Live electric eels that had been shipped to Paris in the 1820s died shortly thereafter because they had been subject to experimentation too soon after the voyage, whereas a specimen brought to Stockholm in 1797 had survived for more than four months.

The British government's efforts were not successful, however. The electric eel that arrived in London in 1838 was brought by a Mr. Porter, exhibited briefly at the Zoological Society and sold to the owners of the Adelaide Gallery. At the gallery, Thomas Bradley took care of it, writing a brief account of its habits for Charlesworth's *Magazine of Natural History*. Bradley admired the fish's ability to adapt since it was "kept in a room daily frequented by multitudes of persons, with only a borrowed light from the skylight, and never feeling the direct rays of the sun; confined in a vessel in which it cannot now stretch itself out at full length; kept warm by water artificially heated; and fed with fish not indigenous to the country it inhabits." He would have been glad to see it in an institution that could take better care of it and thought its life might be extended for years if it were to be moved.[31]

The electric eel remained at the Adelaide Gallery and eventually thrived there. It was in a "very debilitated state" when it arrived in mid-August. Since it wouldn't eat small morsels of meat, worms, frogs, fish, or bread, Bradley tried putting bullock's blood in the water—a London fishmonger's trick for fattening eels—and it soon perked up. It began shocking and eating small fish placed in its tub and was weaned off of the blood. The electric eel was apparently well suited to become a member of fashionable society: "next to its graceful movements will be admired the richness of its covering, which, in appearance, resembles the plush velvet lately introduced for waistcoats; the colour being an admixture of dark puce and brown." Not only lithe and handsome, the electric eel could prove its mettle when the occasion demanded it. Threatened by "a bold life-guardsman," the eel's shock put him "down upon the boards with the clang of cuirass and sword, to the great amusement of the spectators."

The *Gymnotus* tank in the Adelaide Gallery became a popular attraction and a place to rendezvous. Ada Lovelace, who directed people to the gallery to see a Jacquard loom in action (and thus better to understand the work of her friend Charles Babbage), visited the electric eel while there. Babbage took his neighbor Miss Burdett Coutts to see the fish at eight o'clock on a Saturday evening in the summer of 1839 and promised that she could dine with Michael Faraday, too. Faraday wrote to both Babbage and Coutts that day to excuse himself from dinner (he was ill at the time and thought of himself as "a recluse & unsocial"), but he did meet the two of them "to offer his respects to Miss Coutts . . . by the side of the Electric Eel." He later remembered it to her as "a very happy evening."[32]

Michael Faraday had long been interested in electric fish. As a teenager, he made notes about them in a commonplace book that he called his "Philosophical Miscellany," nestled alongside facts that he had read about electricity, galvanism, lightning, oxygen gas, light and color, and other topics. He experimented on his own with electricity and electrochemistry, attended Humphry Davy's last lectures at the Royal Institution in 1812 and thereafter became Davy's assistant through a stroke of good fortune. With Davy in Rome, he wrote to tell his mother that he had seen a torpedo at the Academy del Cimento. Davy and Faraday shared an interest in animal electricity. They worked together on electromagnetism in the 1820s, not always harmoniously, but Faraday was made a member of the Royal Society in 1824 and promoted to the directorship of the Royal Institution laboratory the following year. At the Royal Institution he carried on Davy's tradition of giving informative and entertaining lectures replete with demonstrations. He also initiated a popular Christmas lecture series for youngsters, evidently to his delight as much as theirs. His biographer Alan Hirshfeld writes that Faraday had "a sparkle of youthful wonder that never deserted him" and found the Christmas lectures to be rejuvenating.[33]

In the 1830s Faraday wrote a series of papers on his electrical experiments, helping to lay the foundations for an electromechanical world that would eventually fill with electric motors, generators and transformers. Faraday's investigations would not have been possible with Leyden jars or frictional electric machines. The battery provided steady electric current, and Faraday's devices put that electricity to work. Faraday's own labor was intense, resulting in more than a thousand published pages over fifteen years, detailed and thoroughly cross-referenced. Nearing his fifties by the end of the decade, he suffered from serious memory loss and frequent, debilitating headaches that lasted for weeks at a time. He had trouble finishing sentences and "retreated into a circumscribed universe anchored in research, family, the Sandemanian church, and the Royal Institution." Working with Davy's 2000-cell "Giant Battery" in the poorly-ventilated basement laboratory of the Royal Institution, Faraday would have been exposed to high concentrations of toxic mercury vapor, a substance now thought to cause many of the symptoms he suffered.[34]

"Wonderful as are the laws and phenomena of electricity when made evident to us in inorganic or dead matter," Faraday wrote in his 1839 paper on the electric force of *Gymnotus*, "their interest can bear scarcely any comparison with that which attaches to the same force when connected with the nervous system and with life." Electric fish continued to provide an ideal system in which to explore the vital functions of electricity. At the beginning of the decade, astronomer and chemist John F. W. Herschel had speculated that the brain might be an electric pile that discharged itself regularly, and that electricity communicated along the nerves might, in turn, cause the heart to beat. Herschel thought that the study of electric fishes like the torpedo would clarify how the brain was able to develop electrical signals to transmit through the nervous system.[35]

Faraday finally got a chance to do his own studies of strongly electric fish when the *Gymnotus* residing at the Adelaide Gallery was placed exclusively at his disposal for experimentation. He worked with a pair of assistants and the occasional participation of John Frederic Daniell, Charles Wheatstone, Richard Owen and the physician Robert Bentley Todd. Along with Faraday, Daniell and Wheatstone made up what historian Iwan Rhys Morus calls the "self-proclaimed elite of the London electromagnetic network." The men started work in September 1838 by placing their hands on the fish, receiving strong shocks even though it was "apparently languid." "When one hand grasped the head and the other the tail of the Gymnotus," Owen wrote, "I had painful experience; especially at the wrists, the elbows and across the back. But our distinguished experimenter showed us that the nearer the hands were together within certain limits, the less powerful was the shock."

Allowing the *Gymnotus* to rest a few weeks between experimental trials, Faraday custom-built collectors to conduct the electricity from the submerged fish to apparatus outside its tub. One kind of collector consisted of a fifteen-inch-long copper rod with a one-and-a-half inch disk brazed to one end and a copper cylinder fixed to the other end to form a pistol grip. The rod from the disk upward was insulated from the water by a thick caoutchouc tube. Holding the handle of the probe in a wet hand, the experimenter could gauge the intensity of the shock he felt as he placed the disk in different positions on or near the fish. The other kind of collector was an eight-by-two-and-a-half-inch strip of copper, bent into a saddle shape so it could fit over the back and sides of the *Gymnotus*. A thick copper wire conducted electricity from the saddle out of the tank; the outside of the collector was insulated with caoutchouc to keep the water away from it. When the saddle was lowered down over the fish, the animal could be pressed against a glass plate so that part of its body was "almost as well insulated as if the Gymnotus had been in the air." Raising the creature out of the water in an insulated sling would give the best results, but Faraday was loath to harm it.[36]

By placing a pair of saddle conductors over the electric eel fore and aft, Faraday was able to measure the deflection that its electrical discharge caused in a galvanometer, and to determine that the current flowed from its anterior (positive) to its posterior (negative) end. When he added "a little helix containing twenty-two feet of silked wire wound on a quill" to the circuit, "an annealed steel needle placed in the helix . . . became a magnet." Electricity from the *Gymnotus* could also be used to decompose a solution of potassium iodide. The experimenters were not able to measure the generation of any heat with a thermo-electrometer, however, although they did produce a visible spark using two different mechanisms. Faraday also built a model of the electric eel using a battery of Leyden jars that could be discharged into a tank of water through two large brass balls. "When the battery was strongly charged and discharged, and the hands put into the water near the balls, a shock was felt, much resembling the fish." He noted that experiments with the model "had no pretension to accuracy" but they accorded well with the measured deflection of the galvanometer needle. The human instrument still played a crucial role in experimentation with strongly electric fish, but calibrated subjective experience of the kind Cavendish had pioneered was increasingly being replaced by "self-registering material devices and instrumentation." Skill, as Simon Schaffer put it, was now embodied in scientific instruments.[37]

The fish's external discharge must be accompanied by unknown, internal processes that were opposite and equivalent in force, although they may have taken longer to carry out. However these processes worked, Faraday noted that the *Gymnotus* did not apparently feel its own electric discharge, but it could sense when it had shocked another animal. He speculated that it did this by detecting the mechanical vibrations of its victim's muscular spasms. By carefully testing a large number of points around the body of the electric eel, he came to the conclusion that the fish had no control over the path the current took once it had been discharged. For an electric eel entirely surrounded by water, the lines of circulating electric power would follow the outlines of its body in the same way as lines of force followed the curves of a magnet. If it were surrounded by air instead of water, the fish would only be able to shock with direct contact, instead of through the intervening medium. The ineffectiveness of such a mechanism was underlined by an anecdote that Daniell provided in an 1839 discussion of animal electricity. An electric eel that had been captured to ship to Britain was attacked and killed by a water rat. "Much surprise was excited by this catastrophe; but all wonder ceases when we consider the perfect manner in which the body of the rat is insulated. When he dives beneath the water, not a particle of the liquid adheres to him, and his non-conducting fur, and the air which it contains, clothes him with armour which is perfectly proof against the bolts of his formidable antagonist."[38]

At the end of his article on the *Gymnotus*, Faraday speculated that with closer inspection, the electric organs of fish might "prove to be a species of natural apparatus" that would allow researchers to study the interconversion of electric force and nervous power.

> It is not impossible but that, on passing electricity per force through the organ, a reaction back upon the nervous system belonging to it might take place, and that a restoration, to a greater or smaller degree, of that which the animal expends in the act of exciting a current, might perhaps be effected. We have here the analogy in relation to heat and magnetism. Seebeck taught us how to commute heat into electricity; and Peltier has more lately given us the strict converse of this, and shown us how to convert the electricity into heat . . . Ørsted showed how we were to convert electric into magnetic forces, and I had the delight of adding another member of the full relation, by reacting back again and converting magnetic into electric forces. So perhaps in these organs, where nature has provided the apparatus by means of which the animal can exert and convert nervous into electric force, we may be able, possessing in that point of view a power far beyond that of the fish itself, to re-convert the electric into the nervous force.

It was an idea Faraday's colleague Daniell found to be "of surpassing interest," something perhaps only a physiologist who had mastered everything known about electricity could hope to accomplish. Faraday himself was more circumspect: "We are indeed but upon the threshold of what we may, without presumption, believe man is permitted to know of this matter."[39]

SIX

# Electric Currents

## Detecting Current

In the nineteenth century, electricity profoundly changed human capacities for perception and action. Electricity, magnetism, and light were unified into a single domain of electromagnetic phenomena, and the new technology of telegraphy made it possible to communicate over distances at the speed of light. The telegraph served as a metaphor for nervous systems, and vice versa. Some researchers moved easily between physiology and physics, and many dreamt of ways to wire the gap between thought and electrical impulse. It was becoming clear, however, that nerves could not be conducting electricity directly. As before, the answers would be sought by disassembling animal bodies, but as the century unfolded, these processes were increasingly industrialized. Physiology was becoming a factory discipline.

The concept of animal electricity or galvanism provided a useful way to think about nervous and muscular action for half a century. But after the invention of the voltaic battery, it began to decline "due mainly to the fact that electrophysiological research had outstripped the ingenuity of those who could devise appliances for the detection and measurement of electric current." In the 1820s, the most sensitive current-measuring instrument available was a prepared frog. That was changed by Ørsted's demonstration that a compass needle deflected when it was placed near a current-carrying wire. Used in a new kind of instrument called a galvanometer, the deflection of a compass needle could now indicate the presence, strength, and direction of electric current. This device was to have significant consequences for the understanding of electrophysiological phenomena.[1]

At the Ducal Museum in Florence, the natural philosopher Leopoldo Nobili refined the galvanometer so that it would cancel out the effects of the Earth's magnetic field. He used his improved device to study electric currents in frog preparations. He started by decapitating a frog and skinning it, then placed the trunk in one glass of water and the feet in another. When he connected the two glasses with a wet thread of cotton or asbestos, the muscles of the frog contracted—more strongly if some salt was added to the water beforehand. Nobili used his galvanometer to measure the current created by the frog's body, finding that the needle deflected a few degrees for a plain water setup, and up to 30 degrees for a saltwater one. Furthermore, the galvanometer showed signs of the "courant de la grenouille" for several hours, even though the contractions lasted only a short time.[2]

There were measurements Nobili could make with prepared frogs that he could not replicate with his galvanometer, however. Current from a voltaic pile that was too weak to measure with the instrument could still stimulate vigorous contractions in a prepared frog. Nobili found that "wiring" a pair of prepared frogs nerve to nerve and muscle to muscle canceled out their ability to contract, whereas touching nerve to muscle and vice versa stimulated contractions in both. By building a "voltaic pile" from prepared frogs, placing the legs of one on the trunk of the next, Nobili was able to amplify the current to the point where he could measure it with a galvanometer. (The construction of some of these devices required killing a lot of frogs.) Nobili's student, the natural philosopher and mathematician Carlo Matteucci began studying animal electricity in 1836, taking up where Nobili had left off at the end of the previous decade.[3]

Matteucci first worked extensively with vivisected torpedoes, showing that electrical discharge followed from mechanically stimulating or damaging the fourth lobe of the fish's brain and was not affected by stimulation or ablation of other brain areas, such as the optic lobes, cerebrum, or cerebellum. When he used galvanic electricity to stimulate the fourth lobe, the torpedo discharged an electric shock. When he severed the spinal cord or the trigeminal nerves, the fish would no longer discharge a reflexive shock upon pressure to its body or eyes. The experimental setups that Matteucci created consisted of both inorganic components and animal body parts, as did those of Nobili. To test the regions of the torpedo's nervous system responsible for electric discharge: "I take a live torpedo and expose its brain and the nerve trunks which go to the [electric] organ. I lay down the fish thus prepared on a varnished glass surface, cover the organ with some prepared frogs, and put the two plates of the galvanometer on the fish, one on the back and the other on the belly. I put the two platinum wires of a small trough pile of 15 junctions [a kind of voltaic battery] on one of the nerves of the organ, at a distance of two to three centimeters. Suddenly strong contractions appear in the frogs, and the galvanometer needle deviates by 8 to 10°."

When Matteucci placed the battery wires on tissues of the fish that were near to the prepared frogs (but not part of the torpedo's nervous system), the frog legs did not contract. This showed him that his electric stimulus was not diffusing through the torpedo's body but was acting directly on its nerves. Matteucci also demonstrated that the torpedo was capable of producing a perceptible spark, something that embroiled him in a minor priority dispute with Santi Linari, who was working with torpedoes at the same time.[4]

Matteucci's study of the torpedo convinced him that he needed to examine "all of the facts that could be related to an electric state inherent in animal organs," so he turned his attention to Nobili's "frog current." If any voluntary muscle of a frog were damaged, an electric current could be detected while the muscle was resting, by placing one galvanometer test lead into the wound while the other lead was placed on some intact tissue nearby. The flesh of the wound was always electrically negative with respect to the uninjured surface. This effect—later referred to as the "injury current"—was not limited to frogs; Matteucci cut into the muscles of living pigeons, rabbits, and sheep and found the same phenomenon at work.

Translated into our terms, Matteucci "had revealed the existence of an electrical potential in the resting muscle (our 'resting potential'), which was released when it was injured, and which represented the difference in potential between the inside and the outside of the muscle fibre." He put his understanding to work by using it to create a new kind of electrophysiological instrument: the "galvanoscopic" (or "rheoscopic," i.e., "current seeing") frog. Despite the name, it was simply a prepared frog leg encased in a glass tube so that the sciatic nerve projected from one end like a test lead. When he inserted the nerve into the wounded muscle of another animal, the injury current caused the leg in the tube to twitch. He tried the galvanoscopic frog on fish, eels, and rabbits, concluding that "it is certainly the most sensitive apparatus that we possess, provided it is renewed from time to time." Matteucci also observed that when he placed the galvanoscopic frog against the beating heart of another animal, the leg contracted in synchrony with the heartbeat.[5]

Another instrument Matteucci devised from animal flesh was the electrophysiological pile, a battery that consisted of a series of frogs' legs cut and stacked so that the intact surface of one was in contact with the sectioned surface of the next. Illustrations from his research notes show rows of frog limbs and halves chopped and neatly piled together, often in multiples of three. When he measured the resulting injury current with a galvanometer, the deflection of the needle was proportional to the number of elements (i.e., frog half-thighs) that were used in its construction. These experiments showed that the current he was measuring was truly internal to the animal, since no exogenous elements like wires had been introduced. In design, his pile thus followed from earlier all-animal-tissue preparations made by Humboldt and Nobili.

From frogs, Matteucci went on to create electrophysiological piles from the limbs of "fowls, pigeons, rabbits and dogs." "It is necessary to operate with great rapidity upon these animals," he noted, since "the signs of the current which we are now studying cease very quickly." It was difficult for Matteucci to judge by eye the intensity of the contractions that his various experiments caused in prepared frogs, however, so he had the instrument-maker Breguet build a kymograph ("wave recorder") for him. This was a device that suspended a prepared frog's leg in midair. A very fine wire was hooked at one end to the frog's claw and at the other end to a silken thread winding around a pulley. When the leg contracted, the pulley turned, moving an ivory needle to mark the limit of the contraction. A weight on the leg returned it to the original position after it had finished contracting, but the ivory pointer remained in place until reset.[6]

"One thing is very certain," Matteucci wrote, "which is, that it will ever be impossible to make any advance in the study of the physiological action of the electric current, without having recourse to processes which give the measure of that action." During the 1830s and 1840s Matteucci made important discoveries, and he helped to stimulate a new enthusiasm for electrophysiology. But he also tended to publish overlapping or identical studies in different journals, contradicted himself, came to different conclusions based on the same data, and gave the general impression of being disorganized and a bit bewildered. During Matteucci's career, the center of gravity in electrophysiological research shifted to German-speaking organic physicists, particularly the students of Johannes Müller.[7]

## Organic Physics

While Nobili was measuring the frog current with galvanometers, Johannes Müller was developing a program of research that combined close empirical observation with philosophically rigorous reasoning about the logical connections between phenomena. He studied the optic neuroanatomy of frogs, fish, and other vertebrates, comparing their visual systems with those of invertebrates like insects and snails. He pondered over the specificity of nervous action: what caused one nerve to respond to light while another was sensitive to odor? In 1833, he became professor of anatomy and physiology in Berlin and began writing his *Handbook of Human Physiology*, which would become the leading text in the field for most of the century. In the handbook, he argued that we are not conscious of the attributes of external phenomena per se but rather conscious of "a quality or circumstance of *our nerves* brought about by an external cause." Different sensory organs responded to the same physical stimulus in different ways: a blow to the eye might be perceived as light, a shock to the auditory nerve as a tone. Laura Otis writes that "Müller was always most intrigued by the physiological systems that unified animals. The nervous system interested him so greatly because it created wholeness through communication. The brain, he wrote, 'brings together all the different energies of the different parts of the nervous system into the unity of a self-aware, self-determining individual.'" Müller's outlook reflected, and to some extent rejected, the *Naturphilosophie* of his own training at Bonn University. His students, in turn, would praise him for breaking with the tradition but criticize him for not going far enough.[8]

By the 1840s, the idea that the nerves themselves might be conducting electricity, as a wire did, was losing ground to the position that some hidden, and as yet unexplained, process was at work, and the measurable electricity in nerves was simply one indication of it. The emergence of this new consensus was due in large part to natural philosophers who did not practice medicine and whose work did not require detailed anatomical knowledge. They were "a new type of physiologist, the organic physicist, devoted to the practice and propagation of mechanical reductionism, whereby all physiological phenomena were interpreted by means of the laws of physics and chemistry." Their goal was to unify the sciences, their legacy was to be modern biophysics, and their apparatus would be built from electrical circuits and animal bodies.[9]

Müller was a charismatic teacher who attracted a number of highly gifted students during his career, including some, such as Emil Du Bois-Reymond, who wished to contribute to his program of physiological experimentation. Du Bois-Reymond joined Müller in 1840 and stayed with him until his teacher's death eighteen years later. But by the mid-1830s, Müller had more or less lost interest in doing experimental work in favor of observational and classificatory studies of marine life, and this was to lead to occasional conflict with Du Bois-Reymond, who resented being asked to classify "fossil vermin." Müller had received a copy of Matteucci's essay on animal electricity from Humboldt, and he was suspicious of the claims that Matteucci made. One of the first tasks that Müller gave to Du Bois-Reymond was to determine whether or not nervous impulses were actually electrical in nature. There were reasons to be skeptical: after all, tissues besides nerves could conduct electricity, non-electric stimuli excited nerves, and a nerve that was incapable of sending a sensory or motor signal because it was ligated was still electrically conductive. Du Bois-Reymond was to spend much of a long lifetime studying the electrical activity of nerves and muscles. As his instrumentation became ever more refined, it changed the kinds of questions that he asked and the explanations that he found convincing.[10]

Du Bois-Reymond did much of his life's work with frogs. In his more serious moments he referred to the animal as the "martyr of science" and felt that the suffering that they endured in his experiments must have been "monstrous." Occasionally he was sardonic about studies of animal electricity: "It may be said that wherever frogs were to be found, and where two kinds of metal could be procured everybody was anxious to see the mangled limbs of frogs brought to life in this wonderful way." Regardless of rhetoric, frogs had many advantages to recommend them to experimenters. They were hale and anatomically simple, and not nearly as capable of expressing anthropomorphic pain as, say, a dog—although Du Bois-Reymond noted that frogs in pain did display a "horrible writhing and cooing." They were physically small and usually could be procured daily in large numbers from the rural fringes of cities worldwide.

In the mid-nineteenth century, Berlin's waterways had mostly not been diverted into canals or been subject to straightening or bank reinforcement. When Du Bois-Reymond went to collect frogs himself, it didn't cost him anything but his time, although he complained in 1841 about "walk[ing] across the city for half an hour in the heat with a sack of frogs and ice for nothing" because he couldn't get into the Anatomical Museum to do his work. Sometimes it was less hassle to pay street urchins to collect the animals for him. He turned his room in his parents' apartment home into a frog kennel, sharing it with a hundred animals that would eventually be killed in various experiments, and gaining the nickname "frog doctor" among his neighbors. His dependence on a species that was not (yet) subject to intensive husbandry for laboratory purposes meant that Du Bois-Reymond occasionally suffered from events outside of his control, like losing between a tenth and half of his frogs to epidemics for four winters in a row in the 1840s.[11]

After doing an extensive review of the existing literature on electrophysiology, Du Bois-Reymond built a sensitive galvanometer of his own, winding about a kilometer of wire into a coil with 4,650 turns. This device allowed him to detect electrical activity in pieces of muscle tissue that were so minute that the individual fibers could be counted with a microscope. Normally, the sheath of a muscle or nerve was electrically positive when measured with respect to the core. This difference was reduced or reversed when the tissue was stimulated.

Taking Faraday's description of electrical induction as his inspiration, Du Bois-Reymond theorized that the muscle fiber was electrically similar to a copper cylinder that was zinc-plated on the outside. A bundle of these elements, insulated from one another and suspended in a fluid medium, would maintain a resting difference of electrical potential between inside and outside. If the test leads of a galvanometer were touched to two points on the outside or on the inside of a cylinder, there would be no current. If one were to cut into a cylinder to expose a zinc-copper boundary and place the leads there, however, the galvanometer needle would deflect. When Du Bois-Reymond realized that even the smallest fibers exhibited a current, he replaced the battery-like cylinders with an array of tiny spheres, charged positively on the equator and negatively on the poles. Any stimulus that could polarize the spheres, causing them to flip their orientation, could temporarily induce an electric charge. Du Bois-Reymond dedicated his *Animal Electricity* (1848) to Faraday.[12]

The instruments of Du Bois-Reymond also showed that the same electrical forces were at work in the muscles of human beings as in those of other animals. In the summer of 1846, he attached the test leads of a sensitive galvanometer to two strips of platina, and put each into a container of salt water. He then had his subjects place their left and right index fingers into the containers, wait for the instrument to stabilize, and then flex the muscles in one of their arms while leaving the other slack. Using an electrical technique that Charles Wheatstone had popularized a few years earlier but did not invent (now known as the "Wheatstone bridge"), imbalance between right and left sides of the circuit registered as a deflection of the galvanometer needle, demonstrating human myoelectric potentials for the first time. "Not all of his friends who flexed their biceps could make the galvanometer needle move," Laura Otis writes, "but seventy-nine-year-old Alexander von Humboldt succeeded in May 1849, with Helmholtz and Müller watching." Humboldt became a promoter of Du Bois-Reymond's work, seeing it as a continuation of his own efforts a half a century earlier.[13]

Berlin, the Prussian capital, was rapidly industrializing in the 1840s and 1850s. It was at the center of Werner von Siemens's expanding telegraph network and home to workshops, manufactories and machine shops. "Part of the tension that would arise between Müller and his students in the 1840s can be traced to their eagerness to use local machinists' discoveries in their experiments, while Müller preferred more traditional observations of natural forms." One of Du Bois-Reymond's other mentors was the wealthy physicist Gustav Magnus, who had a superb collection of scientific instruments in his home laboratory and allowed them to be used only by a few carefully chosen students. In December 1845, Du Bois-Reymond met another of Müller's students, Hermann von Helmholtz, in Magnus's lab, and the two became friends. Helmholtz was particularly adept at transferring ideas and techniques from one domain of research to another. His 1847 work on the conservation of energy in thermodynamics drew on quantitative studies of animal heat and muscular contraction to argue that natural forces were interchangeable: whenever energy was apparently lost it had actually been converted into heat instead.[14]

Around the time that he met Du Bois-Reymond, Helmholtz also began collaborating with the telegraph engineer Siemens. In the Prussian army, Siemens had specialized in ballistics and in 1845 he devised an electrical method for measuring the velocity of projectiles based on the length of time before a circuit was broken. The following year Siemens saw a demonstration of the English telegraph and decided to produce telegraphic equipment himself. He was soon putting his measurement expertise to work in his new business. Helmholtz, in turn, saw that the galvanometer-based techniques that Siemens and others were using to measure very brief intervals in ballistics and telegraphy might also be used to study the nervous and sensory systems. The historian Timothy Lenoir states that around this time "Helmholtz conceived of the nervous system as a telegraph . . . he viewed its appendages—sensory organs—as media apparatus: the eye was a photometer; the ear a tuning-fork interrupter with attached resonators."[15]

But Helmholtz was soon to demonstrate that conduction in the nervous system was considerably slower than it was in telegraphic networks. While studying muscular contraction, he built a self-registering device that, like Matteucci's kymograph, included the leg of a frog. In the Helmholtz apparatus, stimulation of the sciatic nerve established an electric circuit that was subsequently broken when the frog's gastrocnemius muscle contracted and lifted a weight. Much to his surprise, Helmholtz found that when he stimulated the nerve at different points, the length of time between stimulus and contraction was measurably different. Since the sciatic nerve of a frog is only a few inches long, if signals were traveling at the velocity of electricity (then thought to be 280,000 miles per second) there should be no measurable difference. Helmholtz realized that he had accidentally built an apparatus that could measure the velocity of nerve conduction. He performed several trials, determining that the nervous impulse was actually traveling about 86.6 feet per second. In other words, nervous conduction took about 17 million times longer than it should if it were nothing more than an electric current. In a public lecture around the same time, Emil Du Bois-Reymond stated that the electric telegraph "was long ago modeled in the animal machine" and that the "kinship" between telegraphic and nervous systems was more than similarity, "an agreement not merely of the effects, but also perhaps of the causes." But Helmholtz's finding obviously raised some complications. Du Bois-Reymond was subsequently to note another difference between telegraph wires and nerves: the latter, "do not, once cut, recover their conducting power when their ends are caused to meet again."[16]

## Universal Currency

Nineteenth-century researchers like Michael Faraday and Hermann von Helmholtz experimented as readily with apparatus that included animal bodies or body parts as they did with purely abiotic equipment. Such hybrid assemblages were nothing new, but they were beginning to make a new kind of sense in the decades before mid-century. Within the span of a few years, at least a dozen natural philosophers in Europe proposed some version of the claim that energy can be converted into various forms but that it can be neither created nor destroyed. The philosopher of science Thomas S. Kuhn famously argued that the "history of science offers no more striking instance of the phenomenon known as simultaneous discovery." Some showed that heat could be converted into mechanical work, and vice versa, and calculated a conversion coefficient. Some argued that force is indestructible, or that the total amount of force is constant, or that there is but one force in the world, one that can manifest as electricity, heat, motion, light, and in other forms. In retrospect, we see all of these as examples of the conservation of energy. The problem that Kuhn raised was to explain which elements led natural philosophers to a new way of viewing the natural world in the two decades before 1850. He identified three: the availability of new processes for converting energy, concern with the engines of the Industrial Revolution, and the influence of *Naturphilosophie*.[17]

Seventeenth-century frictional electric machines demonstrated the conversion of motion to electrostatic charge, and charges produced motion in turn, in the form of attraction and repulsion of light bodies. The steam engines of the eighteenth century (with precedents in classical antiquity) demonstrated the interconversion of heat and motion. After the development of the voltaic pile, however, and even more so after Ørsted's demonstration, an ever-growing series of experimental devices were invented to convert forces of various kinds into one another and especially to convert forces to and from electric currents. The voltaic pile showed that chemical affinity could generate electrical current, which could then be used to electrolyze liquids, thus effecting a chemical reaction. In 1821, Thomas Johann Seebeck discovered that if copper and bismuth wires were joined at the ends and the junction was heated, a nearby compass needle moved. The Seebeck effect thus translated temperature into an electric current that could be visualized when the needle of a galvanometer was deflected by it. A decade or so later, Jean Charles Athanase Peltier showed that electrical current could also absorb heat (or produce cold) in some devices. Kuhn notes that the development of photography after 1827 added still more examples of conversion processes.[18]

Once a force of any sort had been converted into the universal currency of moving electric charges, it could be mobilized in new ways. In 1836, William Fothergill Cooke saw a lecture demonstration in Heidelberg on the use of electrical wires to carry signals. Cooke was an anatomical modeler by trade, visiting Germany to perform dissections. He was inspired by the potential applications of the idea of electrical communication, but he lacked the skills he would need to scale the device up from a proof-of-principle demonstration to a practical, working technology. Back in London, he studied the electricity and magnetism exhibits in the Adelaide Gallery for inspiration, and he soon entered into a collaboration with the natural philosopher Charles Wheatstone, who had been working along similar lines. In the early 1830s, Wheatstone had done a series of experiments at the gallery to measure the velocity of electricity and had shown that it traveled at 280,000 miles per second, which was even faster than contemporary measurements for the speed of light. If electricity could be harnessed to send messages, they would seem to arrive instantaneously, no matter how great the distances involved.[19]

Creating a workable telegraph required powerful electromagnets and a sophisticated knowledge of the factors that could lead to a loss of signal when transmitted over long distances. Wheatstone just happened to have both pieces of the puzzle in hand. He had recently seen a personal demonstration of Joseph Henry's technique for increasing the power of electromagnets by winding their coils in a particular way. And, unlike many of his countrymen, Wheatstone could also read German and was familiar with the experiments that Georg Simon Ohm did in the 1820s on the relationship between electrical force, current, and the properties of the material that a conductor was made from (these studies wouldn't be published in English until 1841). Ohm showed that the electric force divided by the strength of the current was equal to a constant he called "resistance." Resistance, in turn, depended on the length, cross-sectional area and the "resistivity" of the material itself—some materials, like copper, offered very little resistance, others, like paper, very much. An understanding of resistivity and Ohm's law was needed to estimate the resistance of miles of telegraph cable accurately.[20]

Cooke and Wheatstone received a patent for their telegraph in 1837. The technology took off quickly despite contention between the partners, and despite the claims of rival inventors like the American Samuel Morse and London surgeon Edward Davy. Telegraphy was enthusiastically described in the English-language press from the late 1830s onward. Many commentators dwelt on its Godlike potential to annihilate time and distance: "Canst thou send lightnings, that they may go, and say unto thee, 'Here we are?'" (Job 38:35). To John Timbs, who had written several pieces on electrical topics—including an article about the electric eel residing in the Adelaide Gallery—the cables of the metropolitan telegraph network were "the nerves of London." It was a metaphor many Victorians found compelling, and it was used throughout the century by almost everyone who wrote about the telegraphic or nervous systems. Not only did the routing of electrical messages through the telegraph network have something to teach electrophysiologists, but the electrical workings of the animal nervous system contained inspiration for telegraph engineers.[21]

Machines took on a new kind of vitality in organic physics. William Robert Grove—who gave an series of lectures on the "correlation of physical forces" at the London Institution in 1842 and 1843, and who developed new kinds of batteries, including a fuel cell—put it this way: "To understand an ordinary machine—e.g., a watch or a clock—the action of one force alone has to be considered, such as a falling weight, or the reaction of a coiled spring, etc. In a steam-engine, the consideration of another force—heat—has to be added. In a voltaic battery and its effects, the nearest approach man has made to experimental organism, we have chemical action, electricity, magnetism, heat, light, and motion. But in the human body we have all these (and possibly other forces or modes of force of which we are at present ignorant), not acting in one definite direction, but contributing in the most complex manner to sustain that result of combined action which we call life."[22]

The physician and physiologist William Benjamin Carpenter, onetime tutor to the children of Ada Lovelace, was one of those for whom the idea of the voltaic battery as an "experimental organism" took on resonance. Over the course of his career, Carpenter based more and more of his physiology on the conservation of energy. In a letter that he wrote to his mother in 1849, he explained the idea of the interconversion of physical forces and then said that he was led to believe that it also applied to vital forces because electricity and nervous agency were closely related. Electricity could generate both sensation and motion; conversely, the electric fish showed that nerve force could generate electricity.

In an 1875 essay "On the Doctrine of Human Automatism," Carpenter argued that the nerves running from sense organ to spinal trunk and thence to the central sensorium, and those running from motor center to muscle, each served "like a telegraph-wire, to convey a 'molecular motion' (the now fashionable mode of expressing a change of whose nature we really know nothing whatever)." Human nerve tissue came in two forms, as "white" (tubular) and "grey" (vesicular) matter: "the tubular being regarded, like the wires of the electric telegraph, as the *conductor* of nerve-force; whilst the vesicular or ganglionic was considered, like the battery which sends the charge, as the *originator* of nerve-force. We now know that this account of the matter is not strictly true . . . But in a broad, general way, the analogy is sufficiently correct." Elsewhere Carpenter compared the gutta-percha insulation of telegraph wires with the fatty matter that seemed to coat nerves. He was particularly interested in the ability of telegraph clerks to "read the words . . . pass them through their minds, and transfer them to the sending part of the apparatus, just as unconsciously and automatically as Wheatstone's transmitter does." Increasingly, humans and other animals would be seen as comprising and being comprised of networks of communication and control.[23]

The successful spread of telegraph networks required the precise and accurate measurement of electrical phenomena and the establishment of standards. Historians of science have drawn parallels between the engineering workshops where telegraphy was developed and the new laboratories for physics that came into being around the same time and have argued that telegraphy helped to constitute physics in the second half of the nineteenth century. Despite the objections offered by an earlier generation of electricians (including Michael Faraday), the language of quantities and intensities was replaced by resistance, currents, and potential differences, new terms from telegraphy. The theoretical program of unifying forces continued in physics laboratories. Ørsted, Faraday, and others had hypothesized that light might be electrical in nature. The subsequent work of James Clerk Maxwell and his followers showed that this was indeed the case. In 1888, Heinrich Hertz wrote, "The connection between light and electricity is now established . . . In every flame, in every luminous particle, we see an electrical process . . . Thus the domain of electricity extends over the whole of nature. It even affects ourselves intimately: we perceive that we possess an electrical organ—the eye." Light, electricity, and magnetism were all seen to be different aspects of something abstract called an electromagnetic field.[24]

The creation of new energy conversion processes continued apace. In the late 1870s, researchers showed that sound waves could mechanically compress granules of carbon, changing their resistance—their opposition to electric current—and that this could be used to modulate, or "valve," another electrical signal. Sound, in other words, could be translated into a varying electric current. In 1880, brothers Jacques and Pierre Curie discovered that some kinds of crystals generated an electric charge when subject to mechanical strain and, conversely, that they exhibited mechanical movement when an electric current was applied to them. This "piezoelectric" effect made it possible both to build a more sensitive quartz electrometer to measure tiny currents and to translate mechanical force into an electrical signal that could be visualized with a galvanometer. Eventually optical properties, chemical concentrations, viscosity, mass, position, acceleration, shape, surface roughness, and hundreds of other physical stimuli would be visualizable and quantifiable by virtue of being translated first into an electric current. For a time, the twitching needle of a galvanometer became an ubiquitous component of laboratory instruments, an indication that some dimension of physical reality had been sensed by a mechanism that produced an electric response as a result. By the end of the nineteenth century, it was supplemented by the oscillograph, a galvanometer that could display transient or rapidly varying currents.[25]

## Experimental Trials

By its nature, the galvanometer was good for detecting continuous currents, but its response time was very slow because of the inertia of the needle. In ballistic experiments, for example, the falling hammer of a gun established a circuit; as the bullet exited the barrel it cut a wire and broke the circuit. The response of the galvanometer needle was tens of thousands of times slower than the event that it was being used to measure. When an electrical signal was oscillating, the galvanometer could only provide information about the sum of the individual currents and could not follow any rapid changes. This limited its utility for physiological experimentation.

Du Bois-Reymond worked constantly to improve the galvanometer. By wrapping ever larger coils, he could make more sensitive devices: one of his galvanometers had 24,160 turns of wire and was sensitive enough to register electrical activity in nerves that had not been stimulated with electricity. These galvanometers were the best available, and Müller took one of them to Helgoland in the summer of 1846 to study torpedo electric organs. But more sensitive galvanometers did not respond any more quickly than less sensitive ones. So Du Bois-Reymond continued to make use of the galvanoscopic frog when he needed to detect apparently instantaneous currents. The frog's leg would twitch in situations where the galvanometer needle did not. Over the course of the mid-nineteenth-century, the growing availability of electrical measuring instruments that were more sensitive than those made from animal body parts would gradually take frogs and torpedoes out of the day-to-day work of most natural philosophers who were experimenting with electrical physics or chemistry. What could be measured more accurately or reliably was subject to finer control, and inorganic artifacts were not as susceptible to immediate decay as organic ones. The use of apparatus made from parts of living or freshly killed animal bodies became limited primarily to those working in the medical or life sciences.[26]

Du Bois-Reymond considered the properties of the electric organs of strongly electric fish to be of a piece with the more general characteristics of nerve and muscle that he was investigating in frogs. He thought that electrophysiologists ignored electric fish at their peril; unfortunately, Berlin did not have a ready supply of the animals with which to experiment. Theodor Bilharz, an anatomy professor in Cairo, published an anatomical description of the electric catfish in 1857, but living specimens were difficult to obtain even there. The same year, however, enterprising missionaries at the United Presbyterian Mission on the Old Calabar River in Africa arranged for three live electric catfish to be sent to Edinburgh, and they arrived in fine condition despite a shipwreck en route. In Scotland the fish were given to a Professor Goodsir, who "with that total abnegation of self and pure devotion to science which so strongly characterize[d] him, took them to Berlin, to place at the disposal" of Du Bois-Reymond.[27]

Du Bois-Reymond placed the catfish in a glass-sided trough sitting on a slab of slate. Surrounding the trough was a water-filled zinc case, a layer of sawdust, and a wooden exterior. The water in the zinc case was warmed by a copper boiler and gas burner. A black varnished wooden lid allowed air to enter but kept the trough dark, which the *Malapterurus* seemed to prefer. Du Bois-Reymond tried adding earth and water plants to the trough, but they made it more difficult to keep clean and didn't seem to do any good, so he got rid of them. He tested the subjective quality of the catfish's shocks: feeling it to his knuckles if he touched head and tail with his fingertips while it was in the water and above the elbows if he seized it with both hands. "If it is touched with one hand, a pricking sensation is experienced in the skin, a burning one in wounds, and a painful shock is felt in all the joints of the submerged parts." Inspired by Faraday's saddle-shaped collector for the electric eel, Du Bois-Reymond made a gutta-percha shell—"not inaptly compared to the cover of a mummy coffin"—to press down over the catfish and hold it against the glass bottom of its trough so that he could test its electrical abilities. On the inside of the cover, tinfoil linings touched the animal's head and tail, conducting electricity to wires leading up through an insulated handle. The wire leads could be connected directly to a galvanometer, but the absence of a signal might be due to a defective circuit rather than to the lack of an electric shock. Such a setup also required that the experimenter constantly keep an eye on the galvanometer scale so as not to miss anything.[28]

Du Bois-Reymond set about constructing what some scholars might now refer to as a "techno-organic hybrid," although he called it the "frog-alarum" and "frog-interrupter." Wire leads from the gutta-percha collector carried the electric discharge from the catfish through a galvanometer, so that it could be measured. That circuit was broken by a second circuit, which carried electricity from the water of the trough through the leg of a prepared frog. The frog-interrupter was a modification of the apparatus that Helmholtz had used to measure the velocity of nerve conduction. Electricity from the trough also flowed through a second prepared frog's leg via a third circuit. When that leg contracted it pulled a cord that tripped a hammer that rang a bell. The frog-alarum had thus descended (with modifications) from Matteucci's kymograph. Taken together—electric catfish, gutta-percha collector, trough, water, heating apparatus, boiler, gas burner, galvanometer, electrodes, wires, bell, hammer, and a pair of legs from a freshly killed frog—the whole apparatus allowed him to monitor the electrical activity of a fish for many hours at a time. When he put a tench and a loach in the tub with the electric catfish, and a "violent tumult" ensued, and the water became turbid, and the other fish leapt into the air repeatedly to get away from the electric shocks, well, Du Bois-Reymond "should have remained in the dark as to the precise course of events" had not the continuous ringing of his frog-alarum "betrayed clearly enough what took place."[29]

Unfortunately, despite careful handling the electric catfish soon died, and Du Bois-Reymond lamented that his supply of strongly electric fish was so drastically limited. "Nearly every new experiment on the electric organ needs a new preparation. It would have been inconceivable that investigations into nerves and muscles be restricted to two or three frogs."

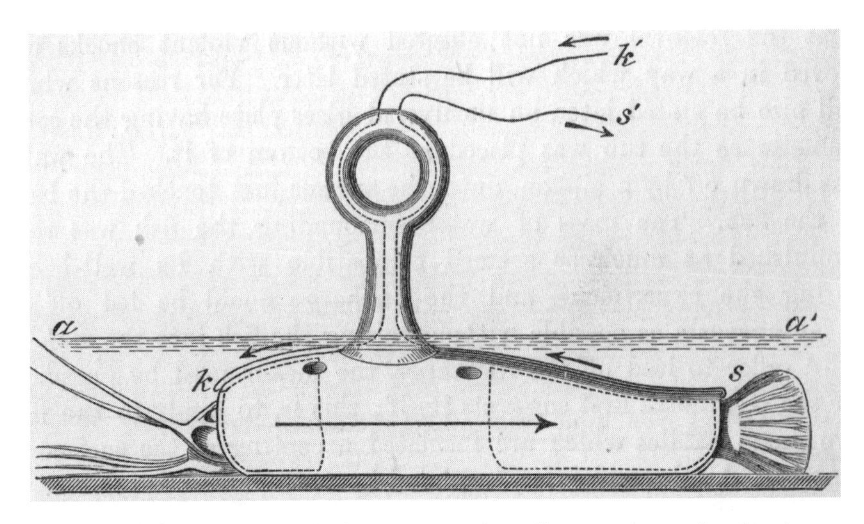


Emil Du Bois-Reymond made this gutta-percha collector to press the electric catfish against the bottom of its tank so that he could study its electrical abilities. From Emil Du Bois-Reymond, "Observations and Experiments on Malapterurus," in *Memoirs on the Physiology of Nerve, of Muscle, and of the Electrical Organ*, ed. John Scott Burdon-Sanderson (Oxford: Clarendon Press, 1887).

The arrival of three more electric catfish at the university in 1858 and two in 1859 did not help much, as only one survived for any length of time. Providing fresh water for the animals' habitat turned out to be difficult. Since electric fish evidently couldn't be brought to Berlin that easily, Du Bois-Reymond decided to mount an experimental expedition to the Llanos of Venezuela, sending his collaborator Carl Sachs to the very place where Humboldt had encountered the *Gymnotus* decades earlier.[30]

After more than a year of preparation, Sachs got on a train from Berlin to Hamburg with a portable laboratory of microscopes, galvanometers, batteries, wire, switches, sheets of copper and zinc, rubber hoses, glassware, thermometer and barometer, weights and a balance, hand tools, syringes, surgical instruments, bottles of chemicals, drugs and poisons (almost a hundred bottles in all), and a lot of other tools and materials that might or might not turn out to be essential. He took a steam ship to South America and loaded the whole production onto five mules to make the journey to riverside. His lab arrived more or less intact by the autumn of 1877, and he set to work following Du Bois-Reymond's carefully laid-out plan of experiments.

As historian Sven Diereg has shown, however, the electrophysiological laboratory of Berlin was no more of an immutable mobile than the *Malapterurus* specimens were. Neither Du Bois-Reymond nor Sachs had realized that the latter would sit through the rainy season drenched and miserable, unable to fish or explore. Many of Du Bois-Reymond's manipulations called for galvanoscopic frogs, but the tree frogs Sachs could capture were too tiny to serve the purpose, and toad legs had the inconvenient characteristic of spontaneously contracting when one did not want them to. Not surprisingly, the *Gymnotus* electric organ turned out to be very different from the one in the catfish. Rather than being innervated by one pair of nerves that could be exposed with a single incision, there were hundreds, each serving only a part of the organ. Sachs tried tying a large bundle of these nerves together to stimulate them simultaneously, but that didn't work at all well. The expedition was an expensive failure. Electrophysiological laboratories with ready access to electric fish were not established until the end of the following decade, in Europe.[31]

Unlike his mentor Johannes Müller, who was a vitalist, Du Bois-Reymond believed that only forces that acted on the rest of the physical world were at work in living organisms. By studying neuromuscular electricity, he was able to attack the notion of a vital force. For the second volume of his *Animal Electricity* (1849), Du Bois-Reymond had drawn an illustration of an electric eel attacking a horse (based, of course, on Humboldt's famous description, which Du Bois-Reymond first encountered via Müller). "In the frontispiece, the tortured horse—whose every muscle is accentuated—suggests life force under attack, suffering from a jolt of animal electricity," Laura Otis writes, "Experimental physiology is zapping natural philosophy." Three decades later, Du Bois-Reymond was realizing that natural history had a way of zapping experimental physiology, too.[32]

## Brain Stimulation

By the 1870s, the electrophysiological frontier was being pushed into the brain. In the middle of the decade, a Liverpool physiologist named Richard Caton thought he might be able to measure electrical activity in the brain with a galvanometer. He tried exposing the cortical surface of the brains of living rabbits and monkeys and touching the test leads to various points. The galvanometer he was using was of a design invented in 1858 by William Thomson (Lord Kelvin) for trans-Atlantic submarine telegraphy: a small silvered glass with a magnetized bit of steel watch-spring cemented to the back was suspended within an electric coil by a thread of cocoon silk. Light was directed onto the mirror so that it reflected on a ruler-like index card. When current passed through the coil and deflected the mirror even slightly, "the beam of light served as a long, weightless lever providing exquisite sensitiveness."

Caton found that feeble currents could be measured on the surface of the brain, or when one of the electrodes was touching the brain and one the skull nearby. The surface of the brain was usually positive, unless it was performing some functional activity, which caused a particular region to become negative. When Caton stimulated a sensory organ, the electrical currents in corresponding brain regions were affected. Shining a light in one of the rabbit's eyes stimulated the portion of its brain that physiologist David Ferrier had shown was related to the movement of its eyelids. Ablation studies, like those done by Matteucci in the torpedo (and by many other researchers in many other organisms), had shown how an animal behaved when parts of its brain were removed. Techniques for doing nondestructive brain measurements and electrical stimulation gave electrophysiologists more precise tools for localization of function and opened up the possibility of doing similar studies with human patients.[33]

Experiments like these still required vivisection. David Ferrier began using electricity to map brain function in living animals in 1873 and was elected to the Royal Society in 1876. The same year, the Cruelty to Animals Act made Ferrier's work illegal, and in 1883 he was charged for performing experiments "calculated to give pain to two monkeys." Ferrier's legal defense was successful, although he was excoriated in the press. Many physiologists still preferred not to use anesthetics as a matter of course—because they tended to interfere with the animals' responses—and thus were widely perceived and portrayed as being inhumane. Some of them feared they would fall behind continental scientists, who did not have to contend with such scruples.

In 1870 at the Anatomical Institute in Berlin, Gustav Theodor Fritsch and Eduard Hitzig removed the skull of an unanesthetized dog, cut into its dura (the outermost meningeal layer surrounding brain and spinal cord), and then electrically stimulated the anterior lobe of one hemisphere with a series of Daniell cells. Groups of muscles on the other side of the animal's body contracted. By moving the electrical probe around the anterior cortex, they were able to create a map of sorts: stimulating a particular area of the brain resulted in the contraction of particular muscles. One controversy of nineteenth-century neuroscience revolved around the question of whether or not the cerebrum was divided into regions according to the functions that it served. Some believed that cognitive activities of various kinds were equally well served by functionally homogenous gray matter. Others, inspired by the work of phrenologists such as Franz Joseph Gall, thought the cerebral cortex to be comprised of functionally specific faculties located in particular regions. Fritsch and Hitzig had provided conclusive evidence for the latter position, and it was their work that had stimulated Ferrier to begin his own studies.[34]

It was in the United States that things were taken about as far as they could be. Cincinnati surgeon Roberts Bartholow took on the care of a "feeble-minded" young Irish woman named Mary Rafferty, who had been working as a domestic. In January 1874, she was admitted to the Good Samaritan Hospital with an ulcer that over the past thirteen months had eroded a portion of her skull, exposing her dura and brain through a hole more than two inches in diameter. In his first report on the case, Bartholow described Rafferty's brain as deeply scarred by injury and by previous surgical incisions that had been made to drain pus. He did not believe that the insertion of fine, insulated needles would cause her harm. Following the work of Farrier, Fritsch, and Hitzig, Bartholow proposed cautiously stimulating different parts of her brain with both faradic (i.e., alternating) and galvanic (direct) currents, and she consented.

Bartholow performed eight trials over four days. Faradic stimulation of the dura resulted in contractions of her arm and leg and the deflection of her head. Using the same stimulus on her posterior lobes led to similar results plus dilation of her pupils and unpleasant tingling in arm and leg. Later tests caused her "great distress; [she] began to cry," followed by spasms of her arm, frothing at the mouth, loss of consciousness, convulsions, and coma. The galvanic tests Bartholow had proposed were abandoned, as "Mary was decidedly worse, she remained in bed, was stupid and incoherent." Further seizures and paralysis were followed by her death. Bartholow's report was published to extensive controversy in the medical community, where he was criticized by Ferrier and the American Medical Association, among others. Bartholow later defended his actions, saying that he believed that Rafferty's death was due to cancer and that "to repeat such experiments with the knowledge we now have that injury will be done by them—although they did not cause the fatal result in my own case—would be in the highest degree criminal."[35]

## Electrophysiological Factories

Although Gustav Fritsch is primarily remembered today for the electrical stimulation work that he did with Hitzig, much of his career was actually devoted to the study of strongly electric fish. After training as a doctor, traveling for a few years in South Africa, and doing a couple of stints as a soldier, Fritsch was appointed in 1867 to be an assistant at the Anatomical Institute in Berlin. The following year he joined the Prussian Solar Eclipse Expedition to Aden—probably as a photographer—and followed that with an archaeological tour of Egypt. At Saqqara, Fritsch photographed the bas-relief in the mastaba of Ty and noted the realistic depictions of fish, including the electric catfish *Malapterurus*. He returned to the Anatomical Institute until 1877, when he was chosen by Emil Du Bois-Reymond to be the head of histology and photography at his new institute for physiology in Berlin. At the time, Du Bois-Reymond's expedition to transplant electrophysiological experimentation to the Orinoco watershed was failing, but he remained determined to find a way to experiment with strongly electric fish. The new institute would allow him and his colleagues to realize that ambition, among others.[36]

Du Bois-Reymond's Physiological Institute was both product and beneficiary of the industrial revolution. It was organized along factory lines, with a division of labor between separate laboratories for physical physiology, chemical physiology, and histology. Sven Diereg argues that Du Bois-Reymond may have adopted the organizational scheme from his close friends Werner von Siemens and Johann Georg Halske, who differentiated their company's telegraph workshops along functional lines in the 1860s. Each of the institute's labs had running water, thanks to a hookup with the city's new urban water network. Each also had a supply of mechanical power, in the form of belt pulleys connected to two gas-driven miniature Otto engines in the cellar. Gas for these engines came via a connection to the city's gasworks. Living animals, subject to vivisection in the labs, were held in a variety of vises, clamps, restraints, and ropes, their voluntary muscles sometimes paralyzed with curare. Since paralyzed animals could not breathe for themselves, their respiration was artificially maintained with bellows driven by the gas-powered engines. "Like the piece on a factory worker's vise, the animal on the dissection table was to be in every sense 'functional,' 'practical,' and 'easily manipulable.'" Du Bois-Reymond's institute was by no means unique. A factory-like approach to physiological research also characterized Carl Ludwig's Physiologische Anstalt in Leipzig (another of Du Bois-Reymond's inspirations) and Ivan Pavlov's institute for experimental medicine in St. Petersburg.[37]

The first living torpedoes arrived at the Berlin Aquarium from Trieste via railroad in the summer of 1881. The director of the aquarium, Otto Hermes, permitted Du Bois-Reymond to make a few studies with the animals as long as they did not interfere with their ability to be exhibited to the public. Once Hermes worked out the details of ensuring a constant supply of live torpedoes (in the months of April, May, September, and October), he offered to provide them to the Physiological Institute on a regular basis. The creatures were kept in the tanks of the Berlin Aquarium, which was only a few minutes away from the institute, until they were needed. Then they could be transferred to the institute, which also had its own aquarium. Fresh water from the city was stored in a tank high in the building and was pumped through the institute's plumbing by a one-horsepower Otto engine. It entered into basins under high pressure and was well oxygenated. Oxygen was necessary not only so that the fish could breathe but also to oxidize their excrement and any particles of food or other dead organic matter that were in the water. "A single Torpedo at one's disposal in a German physiological laboratory," Du Bois-Reymond wrote, "is possibly capable of yielding more for the progress of our science than . . . the Adriatic full of Torpedoes could do at a place where you must first set up a galvanometer, and where, on account of some piece of apparatus left at home or broken in the journey, the most admirable plan of experiment becomes futile." He was speaking from sad experience.[38]

By the mid-1880s, Du Bois-Reymond was able to carry out one experiment after another with electric fish in his laboratory, using organ preparations that were always fresh. Collecting electrical discharges from the living torpedo while it was in the water was trickier than in the case of the electric eel or catfish because both dorsal and ventral surfaces had to be touched while leaving the root of fish's tail free. Instead of using a saddle-shaped collector, he moved the fish into a smaller, cylindrical tank with a circular zinc plate on the bottom connected to one wire. The other wire connected to an arched zinc plate with an insulated handle that could be pressed down on top of the fish. As with his catfish apparatus, a frog-alarum rang a bell every time the torpedo discharged an electric shock. In his experiments with the torpedoes, Du Bois-Reymond was assisted by Gustav Fritsch. Fritsch had taken up the study of electric fish with Du Bois-Reymond's encouragement. He wrote works on their anatomy (1878), contributed sections to the book that Carl Sachs published on his return from South America (1881), and wrote a large, illustrated volume on *Malapterurus* (1887) and another on *Torpedo* (1890). Fritsch made several trips to Africa to study the electric catfish, and sent reports back to Du Bois-Reymond.[39]

In the last quarter of the nineteenth century, electric fish were increasingly incorporated, whole or in part, into hybrid apparatus that was capable of translating mechanical motions into inscriptions via a system of levers. Such devices were long known in meteorology, where they were used to create a continuous record of temperature, barometric pressure, wind direction and force, rainfall, and other environmental variables. Adapted for use with steam engines, graphic indicators could trace the variation in pressure over the operating cycle of the machine. Helmholtz and his friend Carl Ludwig were among those who then introduced similar devices into physiology. Helmholtz created a "frog drawing machine" that recorded the contractions in a prepared frog in much the same way that James Watt recorded steam pressure a half century earlier. Ludwig adapted a mechanism similar to one in a steam indicator to record blood pressure. Continuous curves were traced on a kymograph, now typically a rotating drum covered with smoked paper. Kymographs had been in use for physiological recording since the 1840s, when they began to make visible processes that had unfolded too slowly or too rapidly to be followed by the naked eye. "Once gas engines and transmission belts were

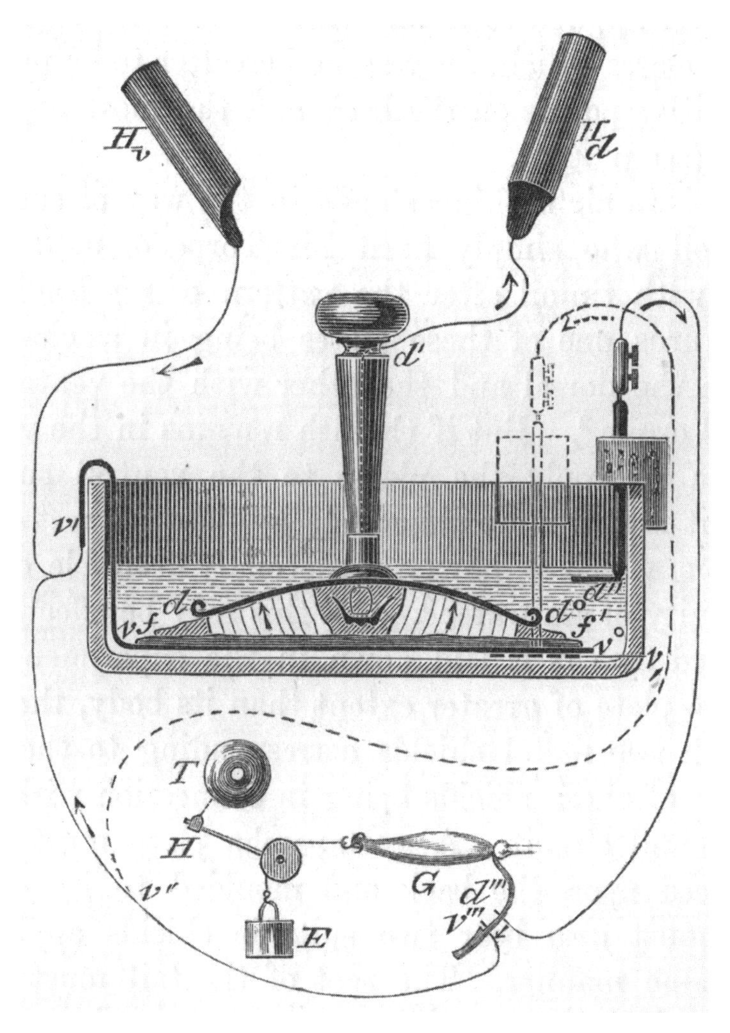


Emil Du Bois-Reymond built elaborate experimental apparatus from living animals, freshly killed ones, and inorganic materials. In the device shown here, a torpedo discharges an electric shock, which triggers contraction in a severed frog's leg, pulling a cable that swings a small hammer against a bell. From Emil Du Bois-Reymond, "Observations and Experiments on Malapterurus," in *Memoirs on the Physiology of Nerve, of Muscle, and of the Electrical Organ*, ed. John Scott Burdon-Sanderson (Oxford: Clarendon Press, 1887).

introduced," Diereg writes, "a laboratory animal could be turned into a curve-producing machine by attaching the kymograph to the belts."[40]

The most ingenious pioneer of this kind of graphical recording was the French physiologist Etienne-Jules Marey. Like Helmholtz and Du Bois-Reymond, he was committed to materialist explanation and was a firm believer in the conservation of energy. Living systems were merely more complex than nonliving ones, not essentially different, and thus the problem was to visualize or trace the flow of force in the "machine animale." In the first part of his career, Marey developed graphical instruments and notations to record heart and respiratory movements, change in temperature and volume, and electrical and neuromuscular events. Later, he turned the camera to the same ends, using high-speed time series "chronophotography" to freeze the motion of running and leaping people, trotting horses, flying birds, bouncing balls, and streams of smoke. His photographic practice was essentially graphical: images collected and displayed data that could not be seen with the unaided eye so that they could be put to use solving scientific problems. "Such instruments both extended the senses of observers allowing them to analyze previously undetected phenomena," Merriley Borrell writes, "and at the same time ostensibly removed the observer from intervening in the measurement of physiological events. As a consequence, researchers could and did claim a new level of objectivity for their science."[41]

In 1876, Marey made the first photographic recording of a bioelectric signal using an instrument known as the capillary electrometer. The physicist Gabriel Lippmann had shown that passing an electric current through a drop of mercury in a solution of potassium bichromate in weak sulphuric acid caused a proportional change in shape in the meniscus. Marey realized that the acid and mercury could be put into a capillary tube and used to block light from a slit lamp, thus creating a magnified shadow image that could then be photographed. Unlike the galvanometer, which gave no indication of variation in an electrical signal during muscular contraction, the capillary electrometer revealed a complete waveform. Marey and Lippmann removed a tortoise heart, hooked it up to the capillary electrometer and photographed the rise and fall of the column of mercury every one twenty-fifth of a second, creating an electrocardiogram. The capillary electrometer was soon turned to recording other bioelectric signals. At Oxford, for example, John Scott Burdon Sanderson used it to demonstrate the electrical activity that accompanied the contraction of the leaves of an insectivorous plant.[42]

During a trip to Naples in the early 1870s, Marey used a prepared frog hooked up to a kymograph to trace the electric organ discharge of living torpedoes. In 1879, he developed another graphical technique for recording shocks from strongly electric fish, this time frog-less. He connected a pair of electrodes running from the fish's electric organ to a telephone receiver and then attached a small pen to the membrane of the ear piece. The electricity from the fish was converted into physical oscillations by the speaker, moving the pen across the surface of a rotating drum. As had been the case with the telegraph, the equipment of telephony was repurposed for electrophysiological experimentation almost immediately. In 1877, Du Bois-Reymond tried connecting a telephone to a galvanoscopic frog. When he called out "Jerk!" into the mouthpiece, the frog's leg twitched in response. (And that was only the year after Alexander Graham Bell's famous request, "Mr. Watson—come here—I want to see you.") Electric signals did not have to be visualized with galvanometer or graphical indicator, they could now be sonified or audified, too. Du Bois-Reymond's students experimented with listening to bioelectric signals. One of them, Julius Bernstein, reported hearing the "rattle" of electrical pulses in the muscles of a prepared frog and the "deep singing tone" of rabbit's leg in tetanus after having been poisoned with strychnine. Touch, sight, hearing: as interfaces were created for each sense, human immersion in an electric world became more complete.[43]

The original Bell telephone did not make weak electric currents audible, but the Siemens & Halske model of the 1880s was more sensitive and was enthusiastically adopted in electrophysiology laboratories. An 1883 newspaper article reported that the sound made by the electric organ discharge of the torpedo was a loud, prolonged moan, whereas that of the electric eel was a sudden blast. The discharge of the common skate sounded like that of the torpedo but was much weaker. Since the mid-1840s, the common skate had been known to possess an organ in its tail that resembled the electric organ of the torpedo. In 1881, Du Bois-Reymond referred to these organs as "pseudo-electrical"; later he decided to call them "incomplete." Burdon Sanderson and Francis Gotch argued, however, that "the organ though small is as perfect in structure and function as that of the Gymnotus or Torpedo." When they attached collectors to the skate and rubbed the skin of its back with a glass rod, the fish created an electrical discharge that could be traced with a galvanoscopic frog and kymograph, could be measured with a galvanometer or capillary electrometer, and could be heard with a telephone.

Around 1890, Hermann and Bernstein developed methods that used the membrane of the telephone earpiece to deflect a beam of light falling on a kymograph drum covered with photographic paper, thus making the audible visible again. Photographic representations of electric signals became common in published electrophysiological studies. Some preferred audification to visualization, however. In the mid-1890s, the Russian physiologist Nikolai E. Wedensky was still using a telephone to listen to torpedoes "sing" at the Marine Biological Station in Arcachon, France, as he believed that telephone and galvanometer each "appear[ed] to be a good witness under certain conditions and a weak one under others."[44]

The self-registering instruments of the nineteenth-century promised to speak in the "language of the phenomena themselves," as Marey put it. As machines, the way they created a trace or an image was itself open to inspection. They appeared to take subjective factors out of science, creating experiences that were more reliable than unmediated ones. As models for animal sensory apparatus, they held out the hope for a scientific understanding of sensation, perception, and cognition. But where eye or ear might be understood to be like a photometer or a tuning-fork interrupter with resonators, or the network of nerves might seem telegraphic or telephonic in organization, some of these new instruments opened up completely new kinds of perception. Rather than merely amplifying existing human senses, they seemed instead to create new kinds of artificial senses. A physiologist armed with a chronophotographic rig was an inhabitant of entirely new worlds.

The historian Robert Brain shows that these scientific instruments were understood in neo-Lamarckian evolutionary terms. In this view, the division of labor among organs and tissues within the body was followed by the creation of tools and artifacts outside the body. These extended and generalized human movements, and instruments, in turn, "extended the functions of human sentience." Marey himself imagined both that the functional movements an organism made, and the external forces it was subject to, left an imprint on its tissues and that these traces could be studied for clues to ontogeny and phylogeny. Underwriting the production and use of self-registering instruments, and the understanding of the living organism in terms of those machines, was a firm belief in the conservation of energy.[45]

SEVEN

# Discovering Electric Worlds

## The "All-or-None" Principle

The experience of animals, including people, is largely constituted by the kinds of things that they are capable of perceiving. Some insects can see in the ultraviolet spectrum; some snakes detect infrared radiation emitted by their prey; bats and dolphins have sonar for echolocation; and so on. As a species, our inclination—right or wrong—has always been to manipulate whatever we can get our hands on, but we need to know what things are there in the first place. We extend our perceptions with instruments that amplify or transduce signals to which we would otherwise be oblivious. We slow down or speed up recordings of events that unfold too quickly or too slowly to follow. Being sensitive to phenomena that occur on distinctly inhuman scales of space or time, we have some hope of controlling them. By the second half of the nineteenth century, people had a good grasp of the workings of electricity on human scales, but it remained difficult to measure electrical events that were very small, rapid, or transient. Around the turn of the century people learned to detect, visualize, and eventually to control these events. The technologies of electricity were supplemented by *electronic* devices, ones that could do useful work with flows of electrons. These would enable researchers to figure out how nerves actually worked, discover the exotic perceptual worlds of weakly electric fish and other animals with electrosensory abilities, and begin to disassemble the neural mechanism's underlying behavior.[1]

As electrometry was refined, it became clear that nerves in action exhibited unusual electrical variations. Emil Du Bois-Reymond had shown that electrical current decreased when a nerve was stimulated, a phenomenon he called the "negative variation." He believed that this negative variation was the signal that was communicated from nerve to muscle to trigger a contraction. Du Bois-Reymond's student Ludimar Hermann pointed out that the negative signal propagating along a nerve fiber might make the external surface of the region ahead become more negative, thus stimulating the nerve and further propagating the signal. In Hermann's model, the nerve consisted of a core and external fluid that were both conductive, separated by an insulating sheet. Hermann's subsequent "local circuit" theory drew on Lord Kelvin's earlier formulation of electrical current to suggest that resting fiber was locally excited by current flowing from the active region.[2]

Neither the galvanometer nor the galvanoscopic frog was sensitive enough to show if the current was merely reduced to zero during the negative variation or if it actually changed sign briefly. In the 1860s, Du Bois-Reymond assigned the problem to his student Julius Bernstein, who also studied with Helmholtz. The inertia of the galvanometer needle prevented the instrument from registering transient currents, so Bernstein adapted a device known as a rheotome ("current slicer") to sample a brief interval of the current at a precise time. By itself, a single sample wouldn't budge the galvanometer needle. But by repeatedly stimulating the nerve, and by always sampling its response at exactly the same time, the minute samples could be summed and the resulting deflection of the needle recorded for a particular point in time. Bernstein modified the rheotome so that he could vary the interval between stimulus and sample, thus creating a "differential rheotome." Using a series of rheotome measurements taken with different temporal offsets, he was able to plot the amplitude of the negative variation in nerve and muscle over time, although this figure did not show the variation of a single nervous impulse, but rather the average magnitude for hundreds of events. Bernstein showed that the excitation of the nerve took about one millisecond. In some of his trials he also observed that the excitation signal did "overshoot" the injury current, but this would be largely forgotten until it was given new emphasis in the 1930s.[3]

As new current-measuring instruments became available, they were used to study electric fish. In 1895, K. Schönlein used a differential rheotome to reconstruct the discharge of an electric eel, and four years later, S. Garten recorded torpedo discharges with a capillary electrometer. Around the turn of the twentieth century, the most important of these new instruments was Willem Einthoven's string galvanometer (1903), which reduced mechanical movements to nearly imperceptible deflections of a quartz fiber and thus lessened the overall inertia of the instrument. Still, none of these instruments was fast and sensitive enough to resolve the succession of individual events that comprised an electric organ discharge.

In the first decade of the twentieth century, electrophysiologists still could not detect or visualize the illusive nerve impulse. These hypothetical impulses "were in three ways minuscule—in space, in time, and in electrical effect—rendering them almost invisible to experimentation." They persisted for only about a thousandth of a second, measured a tenth of a volt, and traveled at 20 to 30 meters per second. "More frustrating was the inability of such instruments even to approach the activity of a single axon in a nerve bundle that physiologists knew contained hundreds or thousands of active fibers, all of whose individual impulses added up to the single tiny blip that he could record," the historian Robert G. Frank wrote. "To create even this, they had to administer an artificial electric shock to the bundle that in no way corresponded to the natural activity of the nervous system. It was much like putting the population of a large village on their telephones, pinching them all simultaneously, and trying to puzzle out the English language from the chorus of profanity recordable on the telephone cable leading out of town."[4]

Not that electrophysiologists lacked theories about what might be going on. In the early 1880s, Svante Arrhenius had shown that some compounds—electrolytes like acids, bases, and salts—dissociated in water into positively and negatively charged ions, allowing the liquid to conduct electric current. In 1889, Arrhenius's former colleague Walther Nernst realized that if there were a semipermeable membrane separating two solutions with different electrolyte concentrations, osmotic pressure would give rise to a difference in electric potential. Nernst developed a set of equations relating the potential difference of galvanic cells to electrolyte concentrations. Around the same time, Wilhelm Ostwald proposed that selective permeability to ions might be responsible for electrical potential across artificial membranes. In the late 1890s, Bernstein drew on all these results from physical chemistry to propose a "membrane theory" of bioelectricity (published in 1902). He imagined the living cell to consist of electrolytic fluid encapsulated by a membrane that was selectively permeable to potassium ions. At rest there would be a potential difference that would only be evident if the cell were damaged (resulting in an injury current). When active, the cell would become more permeable to potassium ions, resulting in the negative variation that had been observed. In a sense, during excitation a nerve cell acted like its membrane had been temporarily or reversibly ruptured.[5]

There was also a growing realization that the electric signal that was propagated along the nerve was independent of the nature or intensity of the stimulus that triggered it. Striated muscle could be induced to contract if it were stimulated by an electrical current above some minimum threshold. Increasing the intensity of the current beyond that point did not increase the strength of the resulting contraction. One possibility was that individual fibers in striated muscle might actually be responding in an "all-or-none" fashion, and that the observed relationship between stimulus and response intensity reflected the number of fibers that had been stimulated rather than the response of each one. Between 1904 and 1908, Keith Lucas demonstrated that this was indeed the case. He isolated a small muscle from the back of the frog and subjected it to a very mild but smoothly increasing electrical current. The muscle gradually flexed, but the function that described the response was staircase-like: each new fiber contributed to flexion in an all-or-none manner. Just before the outbreak of the Great War, Lucas's student Edgar Douglas Adrian showed that the same principle also characterized nerve fibers. Lucas described the effect—in a posthumous publication, as he was killed during the war in an airplane collision—using the metaphor of a train of gunpowder. Once lit by a spark, the fire at each point is hot enough to ignite the powder in front of it, and the train burns until it burns itself out.[6]

Companies that produced commercial electrophysiological equipment helped to mobilize the phenomena that those instruments could demonstrate, and the know-how to make them work. (It took a particular set of instruments, materials, and supplies to cause phenomena to show up reliably in a new place, and that equipment thus served as one kind of immutable mobile.) The physiologist Keith Lucas, for example, was the son of a telegraph engineer and was himself a skilled instrument designer and builder. In the first decade of the twentieth century, Lucas became a director of the Cambridge Scientific Instrument (CSI) Company, which had been co-founded two decades earlier by Horace Darwin (Charles's ninth child). CSI marketed a large number of Lucas's instruments, including devices to assist with capillary electrometry, an adjustable pendulum that turned on current for a very short duration, and "an electromagnetic time signal, which gave accurate regular markings of time intervals on the continuous photographic record of another signal." Without the improvements made by instrument companies like CSI, cutting-edge electrophysiological research would only have been possible in a handful of laboratories. Einthoven's original string galvanometer in Leiden was operated by five people, "weighed several tons, filled an entire room, and needed constant running water to cool the electromagnet." CSI's commercial versions of a decade later were more robust, simpler to operate, and much reduced in scale.[7]

The prewar years also saw continuous invention and improvement in devices that could manipulate electrical currents, which were demonstrated just before the turn of the century to consist of "electrons," negatively charged subatomic particles. The most important of these electron-manipulating devices were vacuum tubes or thermionic valves, as they were called in the US and UK respectively. One of their useful properties was the ability to pass current in one direction only; in 1911, tubes or valves were also shown to be capable of acting as amplifiers. In addition to amplifying small signals, there were new devices to visualize those signals, too. Another instrument manufactured and sold by CSI was the Duddell high-frequency oscillograph, originally developed for power supply engineering. This married an extremely lightweight reflecting galvanometer with a photographic plate that captured a magnified image of the trace of an alternating electric current. Julius Bernstein had suggested that such a device might be used to visualize nerve signals in his *Elektrobiologie* of 1912, written shortly after his retirement, and a few years later K. Fuji used a CSI oscillograph to record the discharges of the Japanese electric ray. The ability to amplify and visualize tiny, rapid, and transient electric currents would soon be felt throughout the field of physiology.[8]

Before the outbreak of the Great War, the Harvard physiologist Alexander Forbes (whose father was the president of the Bell Telephone Company) visited Keith Lucas and Edgar Adrian and conducted some experiments with them. On his return to the United States, Forbes began doing electrical recording in his own lab. As his Allied friends were drawn into the war, he became a volunteer at the radio lab in the Harvard physics department, developing wireless for airplanes and learning how to use the new audion vacuum tubes to amplify weak signals. He moved on to radio development, enlisted in the US Navy, and specialized in radio detection finders. When the war ended, Forbes drew on the manufacturing resources of Bell Telephone to put his new knowledge to use designing a single-stage thermionic amplifier for physiological research.[9]

At Washington University in St. Louis, Herbert Spencer Gasser read Forbes's preliminary report and started working on the amplification of electrophysiological signals. He had two key supporters: Joseph Erlanger, who had already done work on electrical conduction in the heart, and, via a friend, an electrical engineer at Western Electric who designed a three-stage vacuum-tube amplifier capable of gains on the order of twenty-five, fifty, and five thousand. (By comparison, Forbes's single-stage amplifier had a gain around forty or forty-five.) The third stage of amplification proved too much for Gasser and his colleagues, repeatedly burning out the expensive quartz fibers of their string galvanometer whenever their apparatus was hit by a stray current surge.

The newly available low-voltage cathode-ray oscilloscope could work with Gasser and Erlanger's amplifier, however, and they began using one in mid-1921. In the oscilloscope, electric currents deflected a beam of electrons fired through a vacuum, causing an image to appear where they hit a phosphorescent screen. The oscilloscope did not suffer from inertia and was thus able to resolve electric events with one tenth of a millisecond precision. After a single event, however, the screen was too dim to record photographically, so the researchers had to resort to synchronizing the artificial electrical stimulation of a nerve with the sweep of the device to superimpose multiple identical events on screen.[10]

After the war, Adrian corresponded with Forbes about the possibilities of amplifying physiological signals. "The valve sounds an excellent idea," he wrote, ". . . if you don't make it work we shall have to breed a new kind of frog with a large electrical response." Forbes visited Adrian's lab at Cambridge in 1921, and built a single-stage amplifier there. Adrian was excited by the possibilities, but over the next few years he found the instrument limited and was dissatisfied with the way that his research program was going. He contacted Gasser and received instructions that enabled him to build a multistage amplifier. Adrian was running low on funds by that time, did not have access to a cathode ray oscilloscope, and could not afford to burn out string galvanometer wires, so he decided to use the capillary tube electrometer instead.

Although capillary electrometry had fallen out of favor with physiologists by the mid-1920s—and Adrian was one of the last to use the technique—the reasons for its decline were inconsequential after the advent of vacuum tube amplification. The electrometer was fast, durable, foolproof, and its electrical characteristics were a good match for the amplifier. Nevertheless, when Adrian tested the instrument by recording from the nerve of a prepared frog, the results were quite erratic. Eventually he realized that the irregularity he was seeing in the baseline measurement appeared when the leg was suspended and disappeared when it was laid on a glass plate. When the explanation dawned on him, he was "very pleased indeed. A stretched muscle, a muscle hanging under its own weight, ought . . . to be sending sensory impulses up the nerves coming from the muscle spindles, signalling the stretch on the muscle."[11]

Adrian's apparatus, including his choice of biological materials, gave him an unexpected advantage. A string galvanometer would have introduced distortions that could not be corrected, and if he had been able to afford a cathode ray oscilloscope, he would not have been able to photograph traces of single, naturally occurring events in the nervous system. He went on to record the signal that resulted when he pinched the frog's skin, signals from touch and pressure receptors, from pain fibers, from the optic nerve in response to light on the retina. He refined his technique to the point where he could capture the signals from a single nerve ending. Adrian also found that nerves seemed to adapt to stimulus, responding to change rather than to steady state information. (This is why we get used to strong smells or do not notice the noise of a refrigerator motor until it turns off.) In later work, Adrian and his colleagues took to attaching a loudspeaker to their amplifier so they could hear a click each time a neuron fired. This practice of audification would soon become common in electrophysiology labs worldwide.[12]

The vacuum tube amplifier and cathode ray oscilloscope had changed electrophysiology by the 1930s. "A single nerve impulse in a single nerve axon, of invariable height, form, and duration, could be displayed on a screen and captured on film." Furthermore, these nerve impulses were identical to one another, whether heading into the central nervous system from sensory receptors that responded to different kinds of physical stimuli or heading outward to control voluntary or involuntary muscles anywhere in the body. What mattered was not a property of the individual nerve impulse but rather the frequency with which impulses clustered together in trains. In other words, the intensity and time course of sensory stimulus, or of motor effort, were frequency-coded. The adoption of oscillography allowed this new understanding to emerge and provided common ground for British, American, and French researchers. In 1927, Alfred Fessard received funding to buy an oscillograph. He and his colleague David Auger used it to study the electric organs of the torpedo over the next seven or eight years at the University of Bordeaux's Marine Biological Research Station at Arcachon in southwest France. Influenced by Adrian's style of research, Fessard and Auger used the torpedo electric organ as a model nerve center.[13]

The choice of the torpedo as a model system was to play a key role in the gradual overturn of the electrical theory of synaptic transmission. The synapse is the junction between a neuron and another cell. By the late nineteenth century, neuroanatomists were starting to discover that neurons aren't contiguous or in contact—like "the splicing of two telegraph wires" as Santiago Ramón y Cajal put it—but rather separated by small gaps or discontinuities. Researchers from Du Bois-Reymond, in the 1870s, to Edgar Adrian, in the 1930s, believed that the nerve impulse was conducted across the synapse in the form of an electrical current. In 1905, however, Thomas Renton Elliott observed that smooth muscle from which the sympathetic nerves had been removed still responded to adrenin (now known as epinephrine or adrenaline). This led him to hypothesize that this "might be the chemical stimulant liberated on each occasion when the impulse arrives at the periphery." A decade later, Elliott suggested that a chemical substance might also be active during the discharge of the torpedo electric organ. The organ consists of a large array of "prisms." Each prism is composed, in turn, of a stack of about 500 "plates," which are each innervated on the ventral side. When the organ discharges an electric shock, all of the ventral surfaces become negative with respect to the nerveless dorsal sides. Elliott was unable to extract such a substance, however.[14]

By the mid-1930s, a rival, chemical theory of synaptic transmission was becoming increasingly plausible as an alternative to the electric theory. Auger and Fessard, working with both whole torpedo electric organs and individual prisms found that the organ would only discharge if the nerves were electrically or mechanically stimulated. Stimulating the organ itself had no effect. There was also a latency between the application of the stimulus to the nerves, and the ensuing electrical discharge. Both observations supported a chemical, rather than electrical, theory of neurotransmission. At the same time, the German biochemist David Nachmansohn was at the Sorbonne studying the breakdown of acetylcholine by an enzyme then known as "choline-esterase." A student of his, Annette Marnay, found that the hydrolysis of acetylcholine happened at very high rates in torpedoes. Finding common ground, Fessard, Nachmansohn, and Wilhelm Feldberg agreed to collaborate and met for three weeks at Arcachon in the summer of 1939. Together they showed that acetylcholine was necessary and sufficient to trigger the *Torpedo* electric organ to discharge. Neurotransmission thus involved the intervention of a chemical mediator that was released into the synapse by the nerve, interacted with receptors on the membrane of another cell, and was hydrolyzed by an enzyme within a few milliseconds.[15]

## Electroreception and Electronics

An improved understanding of the anatomy and physiology of the electric organs of strongly electric fish contributed to the resolution of the puzzle that had plagued Darwin his whole life: how could such organs arise without apparent precursors? By the late nineteenth century, the electric organs of the strongly electric eel and ray were known to have their embryological origin in metamorphosed muscle. The neurons that innervated them were motor neurons, and the individual electroplates were modified muscle cells. This was also true of the "pseudoelectric" organs possessed by species like marine skates and rays, elephant fishes (mormyrids) from the Nile, and gymnotiform knifefish from South America. The strongly electric catfish *Malapterurus* posed some anatomical puzzles for the scheme, but there was agreement that its electric organ was probably homologous with muscle tissue, too. The fact that strongly electric fish shared habitats with related species who possessed seemingly nonfunctional electric organs called out for some kind of evolutionary explanation.[16]

In later editions of *On the Origin of Species*, Darwin left the question open. Du Bois-Reymond, who had been responsible for the term "pseudoelectric," did not believe that such organs could be adaptive, and his view was maintained by some zoologists well into the 1910s and 1920s. Others, such as his colleague Gustav Fritsch, suggested that the strongly electric organs must be derived from the apparently nonfunctional ones in some unknown way. The Russian physiologist Aleksandr Ivanovich Babuchin disagreed with the characterization of mormyrid electric organs as pseudoelectrical and preferred instead to think of such fish as weakly electric. In 1877, he used two wires to connect a prepared frog to the water in the tank of a pet *Mormyrus* and the leg began twitching. He used this method to test for weak electric discharges in other mormyrid species during trips to Upper Egypt, and other researchers working in Africa soon adopted it. Babuchin also observed that the electric eel and catfish, who live in poorly conducting freshwater habitats, have electric organs that are slender and elongated, whereas the torpedo, whose saltwater habitat is a good conductor, has an electric organ that is shorter and broader. This, too, suggested adaptation.[17]

By the late 1920s, biologists were coming to think in terms of those features of an animal's world that were most perceptually salient to it. At Hamburg University, Jakob von Uexküll portrayed the living animal as the subject of an *Umwelt*, a surrounding world that consisted of, and was limited to, the sum total of features of its environment that it could per-

Weakly electric fish such as *Mormyrus* have the ability to generate and sense electric fields, although this was not suspected until the late nineteenth century. They now serve as model systems for neuroethology. *Mormyrus rume proboscirostris*, from G. A. Boulenger, *Les Poissons du Bassin du Congo* (Bruxelles: État indépendant du Congo, 1901).

ceive and the repertoire of effects it was capable of producing in response. "These two worlds, of perception and production of effects, form one closed unit, the *environment*. The environments . . . are as diverse as the animals themselves." Uexküll's most famous example was a tick, blind and deaf, hanging on a branch and waiting for a passing mammal. "From its entire environment, no stimulus penetrates the tick. But here comes a mammal, which the tick needs for the production of offspring. And now something miraculous happens. Of all the effects emanating from the mammal's body, only three become stimuli, and then only in a certain sequence. From the enormous world surrounding the tick, three stimuli glow like signal lights in the darkness and serve as directional signs that lead the tick surely to its target." The odor of butyric acid in the mammal's sweat signaled the tick to drop from the branch, the warmth of its prey's body triggered it to crawl through hair until it reached bare skin, and the feel of skin sent it burrowing into the flesh in search of blood. That was it for the tick's *Umwelt*, as portrayed by Uexküll.[18]

One of Uexküll's students, Hans Lissmann, became curious about weakly electric mormyrids while completing his doctorate on the behavior of Siamese fighting fish. He read up on the mormyrids and had the sense that something was missing in the account of their electric organs, although nothing would come of this hunch for almost two decades. Lissmann joined the Cambridge University zoology department in 1934 and worked on biomechanics. Around 1949, he visited the aquarium at the

London Zoo and stopped to watch the antics of the African fish *Gymnarchus niloticus*, a mormyriform species closely related to *Mormyrus*. Many fish swim by bending first to one side and then to the other, using their bodies and tail fins to create a backward-moving propulsive wave. Not so for *Gymnarchus*, which holds its slender body rigid while swimming and has a pointed tail rather than the more common caudal fin. To propel itself, the fish sends a ripple along a dorsal fin that runs the length of its body. By reversing the direction of this undulation, *Gymnarchus* is able to swim backward and forward with equal ease. It suddenly struck Lissmann as odd that the fish could apparently avoid obstacles while swimming backward. A few minutes later, he noticed that an electric eel in a nearby tank was capable of the same feat, and something clicked.[19]

Lissmann was able to begin investigating this strange capability when a friend of his who was working in West Africa brought him a *Gymnarchus* as a wedding present in 1950. Putting the fish in a tank in his laboratory, Lissmann placed electrodes in the water, amplified the signal, and displayed it on an oscilloscope. He discovered that the fish was emitting a steady stream of electrical pulses, about 300 per second (300 Hz). His physicist friends could not believe that a fish could be producing the signal and spent days searching his lab for the source of the mysterious hum. Lissmann wondered what would happen if he disrupted the electrical field that the *Gymnarchus* was generating. When he dipped two ends of a U-shaped piece of copper wire into the water, it fled. He tried the same thing with nonconducting wire and got no response from the fish. Lissmann recorded the fish's own signal and played it back in the tank via electrodes; the fish immediately attacked them—the electric analog of placing a mirror in the tank of a Siamese fighting fish. The conclusion was obvious: *Gymnarchus* was not only able to generate an electrical field, it could sense distortions of that field, too.[20]

Although Lissmann is usually credited with the discovery of an electric sense in fish, it might be more accurate to say that he was the first person not to let the discovery get away from him. In the sixteenth century, Montaigne thought that the torpedo might be able to sense its prey with its "miraculous abilities." In the eighteenth, Walsh and Ingenhousz showed that the electric eel could distinguish between open and closed electric circuits. In the nineteenth, Faraday added living and dead matter to the discriminations that the electric eel could apparently make.

R. Wagner hypothesized in 1847 that the organs of the weakly electric skate might be sensitive to electricity, a claim soundly rejected by Müller and Du Bois-Reymond. In 1891, Gustav Fritsch found that mormyrids were surprised by copper electrodes and fled from them. The following decade, P. Kammerer drew attention to the ability of mormyrids in the Cairo Aquarium to swim backward. A few years later, V. Franz proposed that mormyrids might be aware of one another's electrical discharges and respond to them. In 1917, G. H. Parker and A. P. van Heusen found that common catfish could distinguish between glass and iron rods in their tank, even when blindfolded, and thus must be sensitive to the extremely weak galvanic currents created when metal is immersed in water. They tried sending a weak current into the tank through electrodes and found that the catfish would approach the electrodes up to a point. When the current was too high, it would avoid them instead. K. Uzuka replicated their findings with the Asian catfish in 1934. At the beginning of the 1930s, S. Dijkgraaf demonstrated electrosensitivity in dogfish, H. C. Regnart argued that the lateral line was very sensitive to electric current, and W. M. Thornton suggested that blind deep-sea fish might use electric currents for sensing. Both R. T. Cox (1938) and Christopher Coates (1947) suggested that the ability of electric eel to locate their prey in the dark might involve weakly electric discharges.[21]

The second quarter of the twentieth century was an exciting time to be studying sensory physiology. In the late 1930s and early 1940s, scientists had demonstrated that bats could both emit and perceive ultrasonic sounds, and that they were using these sounds to navigate in the dark and to locate their prey. This research on bats led to the discovery, in the late 1940s, that dolphins were also capable of echolocation. At the same time, the adoption of vacuum tube amplifiers and cathode ray oscilloscopes continued to give electrophysiologists new insight into the workings of the nervous system. The nerve impulse was now seen to be the product of concentration gradients of sodium and potassium ions, which rapidly depolarized and then repolarized the cell membrane of the nerve. "Neurons could be shown to modulate each other's activities, millisecond by millisecond, and thereby to create the massive complexity of nervous activity." In laboratories like the one associated with the Cambridge zoology department, the equipment and expertise necessary for working with tiny, rapidly fluctuating electric signals were already in place and could be put to many new uses, including explorations in sensory and perceptual physiology.[22]

Over the same period, the word "electronics" began to be used to describe the design of devices that could manipulate flows of electrons. These flows could be specified in terms of voltage (*V*) and current (*I*). (One analogy that is often made is with water flowing through a pipe: voltage is analogous to water pressure, current to a measure of the amount of water flowing past a given point in a unit of time.) As Paul Horowitz and Winfield Hill wrote in their classic *Art of Electronics*, "Crudely speaking, the name of the game is to make and use gadgets that have interesting and useful *I*-versus-*V* characteristics." In a resistor, a conductor that has a known resistance, the current is proportional to the voltage. Devices that are sensitive to physical phenomena like light, sound, and temperature often work by providing a resistance that varies with the magnitude of the stimulus. In a diode, current can only flow in one direction. In a capacitor—descendant of the Leyden jar—the current is proportional to the rate of change of the voltage. And so on.

As each new electronic component was developed and "black boxed," it could be put to use in any number of ways. Pairs of resistors could be used to add, multiply, and divide voltages. Resistors and capacitors could be combined to create circuits to integrate and differentiate. Operational amplifiers, originally built from vacuum tubes, made it possible to take logarithms and antilogs, count events, compare voltages and change state when one exceeded another, and solve differential equations. Electronic circuits, in other words, could compute, even before the development of digital computers. Electrical technologies constituted a new world for people to explore and inhabit; electronic ones allowed them to populate that world with artificial entities that would perceive, think, and act in ever more sophisticated ways.[23]

A range of technically minded individuals, mostly men, gained experience with radio during the wars, or as hobbyists in the interwar years, and amateur radio enthusiasts helped to spread basic electronic skills and equipment far and wide. World War II greatly increased the demand for vacuum tubes to be used in communication devices of various sorts, and for batteries for flashlights and "handie-talkie" radios. At the beginning of the war these batteries were unreliable, short-lived, and made from components that were relatively difficult to obtain. When Samuel Ruben developed a small mercury battery in 1942 that was longer lasting and more reliable, the company that he licensed to manufacture them was "literally *forced* into the battery business" to meet wartime need.

Its origin as an artificial electric organ forgotten, the battery allowed electronic devices to become miniaturized, portable, and ubiquitous. Wearable or portable hearing aids, electric wristwatches, and radios began to change the sensory, perceptual, and communicative niches of the people who used them. In December 1947, William Shockley, John Bardeen, and Walter Brattain demonstrated the transistor, a solid-state amplifier that could be made much smaller and more robust than an equivalent vacuum tube. By the middle of the following decade, transistors were being produced in high enough quantities to be used in consumer devices like radio receivers, and by the late 1950s entire circuits containing many transistors could be integrated into a single, solid-state package.[24]

Hans Lissmann's own work was facilitated by access to these new technologies and to others. In the 1990s he wrote, "Before 1950, work with these animals was hampered by practical matters . . . electric fish, especially those from tropical countries, remained difficult to obtain. The upsurge of interest followed greater availability when air travel and air freight came within reach, and polythene bags replaced cumbersome kerosene cans as transport containers for fish. A further great help was the development of portable electronic equipment which became available for field and laboratory work." Lissmann published his initial findings in a short note in *Nature* in 1951. By that time his *Gymnarchus* specimen had died, and he was invited to travel to Ghana to collect more. While searching for the fish in the Volta River and its tributaries, Lissmann noted that the water was "extremely turbid" during the dry season, full of fine particles that took days to settle out when it was put in an aquarium. Rainfall led to soil erosion and made the fish's natural habitat even more murky. "Under such conditions," Lissmann argued, "the eyes of the fish would appear of little use and this may be expected to favour the evolution of alternative sensory mechanisms." In fact, Lissmann's own eyes were of little use in locating the fish for the same reason: he couldn't see through the muddy water. Without evolved mechanisms to fall back on, he used battery-operated technology instead, amplifying the signals of a pair of electrodes suspended in the river. He confirmed the uniform 300 Hz hum he had discovered in the laboratory and detected several other patterns of electrical discharge as he thumped on the bottom of his boat, threw stones in the water, and netted specimens to bring back to Cambridge. On the Black Volta, his equipment was occasionally subject to extremely powerful and brief blasts that he attributed to the electric catfish *Malapterurus electricus*, which was the only strongly electric fish to be found there.[25]

From observations of his *Gymnarchus* specimens, Lissmann already knew that they were quite agile in pursuit of their prey and exhibited "marked cannibalistic tendencies." What role did their electroreceptive sense play in their behavior? Back in the lab he began systematic experimentation. The fish responded when he moved a small bar magnet outside of its tank. He found that he could fence the animal in by putting a wire rectangle made from bare copper on the bottom of its tank and then placing the fish within the boundaries of the wire. As it approached each side of the rectangle it would shy away, detecting the presence of the metal but not realizing that its path was actually clear.

Lissmann found that two specimens of the fish could detect one another at a considerable distance. Keeping in mind their willingness to eat each other, and not wanting to lose any of the precious fish, he devised a clever apparatus to show that they were using electroreception to locate one another. He placed recording electrodes in the center of the tank, amplified the fish's own electrical signal, and then used a switch to route that signal to one of six pairs of electrodes placed at the edge of the tank. As he switched the signal from one place to the next, the fish would attack the live electrodes. He got similar results by feeding a range of different electrical signals besides the fish's own into the tank. "This experiment may indicate the general anti-social tendencies in *Gymnarchus*," he noted dryly, "but it may be expected under certain circumstances, e.g., in the breeding season, these tendencies are not indiscriminate." That, of course, would not be heritable.[26]

## Imprimence

*Gymnarchus* and other weakly electric fish were evidently able to locate objects, but it was not clear how they were doing it. In 1953, the head of the Cambridge zoology department made the somewhat unconventional but inspired decision to hire a physicist named Ken Machin, whose earlier work had been in radio astronomy. The hope was that Machin would find biophysical problems that he could solve in collaboration with the Cambridge zoologists. He did not disappoint, setting to work with Lissmann immediately.

The papers that Lissmann and Machin wrote together were much more quantitative than Lissmann's earlier work, made use of sophisticated equipment and took advantage of Machin's specialized training in electromagnetic theory. Together they determined that *Gymnarchus* was tens to hundreds of thousands of times more sensitive to electric currents than other fish like minnow, carp, or goldfish. This meant that *Gymnarchus* was able to detect the extremely weak electrical current generated by immersing a single metallic electrode in the water. Using equations developed by James Clerk Maxwell in the nineteenth century, Machin modeled the electrical field around the fish in the absence of any perturbations and then determined the change of potential due to the introduction of an object. One of the terms of this model was a measure Lissmann and Machin called "imprimence": a quantitative measure of the imprint of an object on an electric field. By using this term, they hoped to avoid "the subjective implications of such words as 'electrical perceptibility' or 'visibility.'" They were able to show that their model system responded more strongly to nearby objects than distant ones and could be used to determine which part of the fish's body was nearest to the object.[27]

Following in the footsteps of experimenters like Cavendish, Machin also built a physical model of the fish out of acrylic. He placed pick-up electrodes around the outside surface of the artificial fish and suspended it in a shallow tank with voltage flowing through an electrically conductive fluid. He then placed a large insulating object near the probe electrodes, and made one delicate measurement after another until he was able to locate it. Machin's office mate of the time later remembered, "There was a feeling of rivalry between Machin and the fish; which could make the finer distinctions? The fish won hands down." But then, the fish's ability to discriminate objects in weak electric fields had been shaped by millions of years of natural selection. Machin's physical and mathematical models served as proof of principle for the feasibility of detecting and locating objects based on their imprimence alone.[28]

Lissmann and Machin also devised a series of behavioral experiments to confirm that *Gymnarchus* was able to detect and locate objects based solely on their imprimence. Their idea was to create two objects that differed in imprimence but that should appear identical to all other known sensory systems. If the fish was able to distinguish the two objects, Lissmann and Machin should be able to use that fact to train it to associate one of the objects with food or with punishment. They eventually settled on a pair of porous, opaque ceramic pots sealed with cork and rubber. By themselves, the pots had a very low imprimence. When placed in the tank that Machin had used to test his artificial fish, they barely disturbed the uniform electrical field. The experimenters were then free to fill the pots with solutions of different chemical composition and electrical conductivity, to see which the fish could distinguish. *Gymnarchus* turned out to be capable of making a range of increasingly subtle distinctions based on imprimence: between air and paraffin wax, aquarium water and wax, and even between aquarium water and a solution of 75 percent aquarium water plus 25 percent distilled water. (Distilled water has fewer charge-carrying ions, and thus a lower conductivity.) They were also able to show that *Gymnarchus* was using electrical conductivity and not chemical composition because the fish couldn't distinguish among containers filled with aquarium water, distilled water plus potassium chloride, or distilled water plus acetic acid, as long as the three solutions had similar conductivity.[29]

Machin and Lissmann carried out a subsequent series of experiments to determine how the electrical receptors of *Gymnarchus* operated. They knew from Lissmann's original experiments that the fish emitted about 300 electrical pulses per second, each one a millisecond long. Based on their mathematical and physical models, they knew that the imprimence of objects in the fish's nearby environment would affect the amplitude of these pulses when they were measured by the fish's electric receptors. Their question was how that information was being received by the fish, converted back into a form that could contribute to its survival. There were two possibilities, both already familiar from the world of signal processing and widely used in radio and other forms of telecommunication.

The first explanation, favored by Machin and Lissmann, was that the fish were making use of a frequency-modulation system. In this account, the electric receptors were measuring the mean value of the incoming signal over a fixed window of time. It wouldn't matter if the intensity, the duration, or the frequency of electrical pulses were altered, as long as the product (intensity times duration times frequency) remained the same. The alternate explanation was that the fish had a phase-modulation mechanism. Each time that it emitted an electrical pulse, the hypothesis was that it also sent a time-delayed pulse through its own nervous system. Since the time delay was posited to be proportional to the amplitude of the emitted pulse, by comparing internal and external signals it could recover the amplitude of the received pulse. Machin and Lissmann noted that such a system would be sensitive to the intensity of incoming signals but not to their duration, and they were able to show that *Gymnarchus* was sensitive to both. Thus, their experiments favored the frequency-modulation mechanism.[30]

Hans Lissmann was well aware that the electric organs of strongly electric fish had seemed to Darwin to pose a "special difficulty" for his account of evolutionary development through a series of transitional stages. The discovery that weakly electric fish could create a field around themselves and use it to sense their environment provided Lissmann with a potential answer to the evolutionary puzzles that Darwin had posed.

The explanation started with another sensory organ of fish, the lateral line. In 1935, Christopher W. Coates, head curator of the New York Aquarium, became interested in the electric eel. He put the eel in a hard rubber trough, inserted a pair of electrical contacts through the side, and used the animal's power to light a neon lamp. Repeated three times daily for a public audience, this demonstration quickly became the high point of a visit to the aquarium. Coates continued to study electric eels and published a paper in the *Atlantic Monthly* in 1947 that suggested that the fish had a radarlike apparatus in the enlarged pores of its lateral line that emitted electromagnetic waves and measured the time delay of their reflection from the surroundings. Lissmann rejected this idea for both physical and physiological reasons, but he did think the lateral line had played an important role in the evolutionary development of electric organs.[31]

Physiologists generally agreed that the fish's lateral line served as a mechanical receptor, sensitive to vibration, pressure, and the movement of small water currents. Lissmann hypothesized that the lateral lines of the first ancestors of electric fish were sensitive to mechanical disturbances but were also incidentally sensitive to electric stimuli. Even without being able to generate their own electrical discharges, these fish would have had the advantage of being able to sense the electrical fields created by the muscular movements of nearby prey or predators. This would be particularly useful in turbid water. They would also have been able to sense the electrical fields created by the movement of their own muscles. "However," Lissmann wrote, "this specialization, once it has begun, does not only involve the organs of this one reflex arc. New locomotory mechanisms, new epidermal adaptations to electrical conductivity, etc., may follow, so that finally it can be imagined that such a fish, living in a private, electric world of its own, receives a variety of information through sense organs distributed over the surface of its body which may be likened to an 'electro-receptive retina.'" His choice to compare these electric organs with the retina was not accidental, for Darwin's other prime example of an organ that posed difficulties for his theory was the eye, which was also too complicated to have evolved in a single leap. Once weakly electric fish had evolved, there could be selective pressure to increase the extent of the organ that generated electricity, thus increasing the voltage of the discharge, until "a stage will be reached when 'objects,' which previously could only be located, can now be stunned and swallowed."[32]

## Wave-Type and Pulse-Type Fish

Passive electroreception is now believed to have emerged about 500 million years ago. Many vertebrates have dermal electroreceptors that are exquisitely sensitive to very low frequency electric fields, suggesting that this was an ancestral trait that allowed animals to locate prey and perhaps to orient themselves. "Then as now, the natural stimuli would be local dipole fields generated by other animals and more uniform and large-scale fields created by ocean currents, the animal's own movements in the earth's magnetic field, and by electrochemical sources such as temperature or salinity transitions." These electrosensory organs appear in jawless fishes like lampreys, in cartilaginous fishes (sharks, skates, and rays), and in bony fish. There are also two orders of amphibians containing electroreceptive members, and even three species of electroreceptive mammals, egg-laying monotremes including the platypus, echidna, and close relatives. All of these animals can sense electric fields even if they cannot produce them. When adaptive radiation led to the bony fish, electroreception was lost, and then subsequently reevolved twice in the group. When vertebrates became terrestrial, it was also lost, re-evolving in the monotremes. In the 1920s, researchers attributed the ability of the platypus to catch prey underwater at night with its eyes, ears, and nostrils closed to a "sixth sense." It was not until the 1980s that they began to suspect the egg-laying mammals might actually be electroreceptive: the platypus and echidna will, for example, dig up live batteries but ignore dead ones.[33]

Much of the early work on passive electrosensory systems was done by Ad Kalmijn. In the 1960s, he returned to an observation that had been published thirty years earlier: sharks were sensitive to rusty steel wire. Even when blindfolded they would orient their heads to avoid a wire moving toward them from several centimeters away. To test sharks' ability to sense electric fields, Kalmijn permanently implanted electrodes in them that allowed their own heart rates to be recorded on an electrocardiogram. Electric fields with a voltage gradient as low as 0.1 microvolt per centimeter caused a temporary, reflexive heart rate deceleration when the sharks experienced it. Heart rate deceleration is a part of the orienting reflex and is one measure of an organism's ability to perceive a particular stimulus. In naturalistic settings, sharks were observed to accurately attack live prey buried in the sand. In the laboratory, Kalmijn ruled out alternative explanations for apparent electrosensory ability with several experiments. In one, a pair of live electrodes and a control pair of electrodes (without electric current) were fitted to either side of a polyvinyl plate on the bottom of the test area. Liquefied herring was squirted up onto the plate from a tube in the center. Dogfish regularly attacked the live electrodes, rather than the herring or the control electrodes, indicating that they oriented toward the weak electric current rather than the source of the odor or the electrodes themselves. In the 1980s it was shown that common bony fish "leak enough current from skin, gills, mouth, etc to cause millivolt potentials . . . adequate for a shark to detect at 50 cm."[34]

A small subset of electroreceptive fish—about 1.3 percent of all fish species—are also electrogenic. In other words, they possess electric organs consisting of muscle cells that have been modified into electrocytes, and as a result they are capable of generating weak or strong electric discharges. There are few species that are intermediate between those with electric specializations and those without. This suggests that, unlike dermal electroreceptors, specialized electric organs were not a feature of ancestral vertebrates but rather evolved independently. This seems to have happened at least six times. Electric organs emerged twice amongst the cartilaginous fishes, appearing in about thirty-eight species of torpedoes and more than two hundred species of skates. They also appeared four separate times in ray-finned fishes. They are found in all of the African mormyriform fish (about 200 species), in all of the South American gymnotiforms (about 130 species), in several families of siluriform catfish (including *Malapterurus*) and in one family of perciform stargazers. The stargazer is somewhat anomalous, in that it appears to lack specialized electroreceptors, and its electric organ discharge "although temporally associated with prey capture . . . would seem to be too low in power even to disorient a prey and too late in time to be a sensory guide." Lesioning the stargazer's electric nerve seems to have no effect on its ability to find and consume its prey.[35]

Since the 1950s, electric fish have become increasingly popular experimental subjects, with dozens of laboratories now working on a wide range of research questions using various species. The electric eel's ability to regenerate large amounts of lost tissue, including spinal cord, holds promise for human sufferers of spinal cord injury. In the 1970s, the electric organs of both *Electrophorus* and *Torpedo* were used as sources of acetylcholine receptor in unraveling the cause of myasthenia gravis (an effort that also required cobra venom, paralyzed rabbits, and lots of lab rats). Researchers now hope to be able to use an understanding of *Electrophorus* genetics to learn how to convert normal cells into electrocytes, cells that can generate electricity. They would like to use these electrocytes in vivo to treat medical conditions including Parkinson's disease, epilepsy, muscular dystrophy, and cardiac arrhythmia, all of which involve dysfunction of electrically excitable membranes. Nonmedical applications for electrocytes include the possibility of genetically engineering "single-celled photosynthetic eukaryotes . . . to convert sunlight directly into usable electric currents." Such cells might even be engineered to self-organize into electrically active filaments, biofilms, or "biobatteries."[36]

More basic science focuses on mechanisms of electroreception and electrogenesis: how they develop ontogenetically and phylogenetically, how they are supported and controlled by neuroanatomy, their degree of plasticity, their physiology, and their role in behavior and communication. In the late 1950s, F. P. Möhres showed that *Gnathonemus,* a mormyriform fish, dramatically altered the frequency of its electric pulses in response to conspecifics, thus indicating that electric organ discharges could be used for social communication as well as object detection and location. Over the following decade, research focused on determining how an electrogenic fish distinguished its own electric organ discharge (EOD) from those of its neighbors, how EODs were coded, and what signal repertoire might be understood by any individual member of a given species. Researchers also looked at the neural mechanisms responsible for generating new signals.

By monitoring weakly electric fish in their natural habitats, it soon became clear that there are "wave fish" and "pulse fish." In the former, including *Gymnarchus* and *Apteronotus*, the EOD consists of a series of rapid pulses, each of which lasts about as long as the interval between pulses. When sonified, we perceive the resulting signal as a hum or continuous tone, ranging from a few hundred cycles per second to as high as 1800 Hz. In the latter, including *Gymnotus* and *Gnathonemus*, the number of electric pulses is typically around fifty or fewer per second, and the inter-EOD intervals are much longer than the pulses themselves. Marine skates have a pulse-type electric discharge. The strongly electric catfish, torpedo, and electric eel are classified as having an "intermittent" EOD. Strongly electric intermittent, weakly electric wave-type, and weakly electric pulse-type fish are all found in both Africa and South America.[37]

## Neuroethology

Electric fish have served as one of the best model systems for neuroethology, the study of the neural mechanisms that produce naturally occurring behavioral adaptations. Bioelectricity is relatively convenient to work with in the laboratory and in the field, thanks to continual progress in microelectronics and computation. The large number of freshwater species in South America and Africa provide many opportunities for comparative work, and electrosensing has much in common anatomically and physiologically with audition. In the mid-twentieth century, founders of ethology like Niko Tinbergen and Konrad Lorenz focused attention on "the natural stimuli that elicit biologically important behaviors such as feeding, fleeing, courtship, and fighting" and "the spatiotemporal structure of ensuing action patterns (pursuit, biting, threatening, calling, etc.)." But ethological descriptions of behavior tend to be relatively "thick," to be situated in natural or naturalistic contexts, and to draw on multiple internal and external causes such as the animal's motivation, physiology, and development. This holism makes it much more difficult to trace relevant neural pathways by isolating them from the organism as a whole and its surrounding environment. Instead, simplified physical and conceptual models are crucial in neuroethological explanations of behavior.[38]

Technological systems have long provided a source of metaphors and models for understanding neural activity in living organisms. We have already seen pneumatic, hydraulic, vibrational, electric, telegraphic, and telephonic examples. By the mid-twentieth century, the use of electronic, robotic, and computational analogies was on the rise. In the abstract framework of cybernetics, humans, animals, and machines could all be analyzed in terms of communication and feedback control. The mathematician Norbert Wiener drew on a well-established metaphor when he compared neurons to telegraph repeaters in his 1948 *Cybernetics*. To the neurophysiologist Warren McCulloch, reflexes were analogous to steam engine governors, and computational mechanisms might be built from simplified, artificial units that mimicked biological neurons. These artificial "neural nets" would become a thriving interdisciplinary research topic later in the century. The neurophysiologist W. Grey Walter created models of the nerve and of reflex activity from analog electronic components (resistors, capacitors, and tubes) and built wheeled, phototactic robot "tortoises" (*Machina speculatrix*) that exhibited surprisingly complex interactions when placed together. Electronic analogs of neurons were created by many researchers. L. D. Harmon's "neuromime," for example, could be wired in such a way as to inhibit its own firing or that of other model neurons. More complicated circuits, like amplifiers and oscillators, were also used to model neural activity at various scales.[39]

Among ethologists, Konrad Lorenz often used technical systems as metaphors when describing ethological mechanisms. In an essay of 1950, he compared instinctive action to a hydromechanical mechanism where "action specific energy" was modeled by a flow of liquid, inhibition by a spring-loaded valve resisting the flow, the intensity of stimuli by a jet of water through the system, and the activity that was elicited in response as the outlet of liquid from a trough with multiple openings. In his *Foundations of Ethology* (1981), Lorenz argued that an organ like the eye was far simpler than "the minimum degree of complication which even a man-made electronic model would have to possess in order to simulate, in the simplest possible manner" a behavior pattern involving social communication. He pointed to the emergent behavior of Grey Walter's cybernetic tortoises to make his point.

Unlike the cyberneticists, however, Lorenz did not explicitly model the nervous system with electronics or study the behavior of artificial systems. For him, electronics provided a metaphor for how "during evolution, new systemic properties often arise through the integration of subsystems which, up to that moment, had been functioning independently of one another." A circuit built from a battery, a resistor, and a capacitor displays relatively simple behavior, as does one where the capacitor is replaced by another electronic component, an inductor. A circuit built from a battery, resistor, capacitor, and inductor, however, exhibits a completely new systemic behavior that was not displayed by any of the components before integration: oscillation. Electronics had come a long way in a short time, from crude and bloody models of the torpedo's electric organ to a world filling with a profusion of microscopic specks of silicon, germanium, metals, and other inorganic materials capable of increasingly sophisticated feats of lifelike behavior.[40]

Walter Heiligenberg, a student of Konrad Lorenz and a pioneer of computational neuroethology, provided one of the first detailed neuroethological descriptions of a complicated behavior in a vertebrate species, tracing a path from sensory input through neural mechanisms to motor output. The behavior that Heiligenberg studied was the jamming avoidance response (JAR) of the gymnotiform *Eigenmannia*. Both African mormyriforms and South American gymnotiforms include weakly electric species that can generate wave- and pulse-type discharges for electrolocation and communication. In addition to the dermal electroreceptors, which are sensitive to low-frequency electric fields, these fish also have electroreceptive tuberous organs "which are most sensitive in the spectral range of the animal's own EOD," acting essentially as high-frequency band pass filters. Fish with wave-type EODs, of which *Eigenmannia* is one, tend to slowly alter the fundamental frequency of their electric discharges over a period of weeks or months, raising the question of how the tuberous receptors stay tuned to the fish's own EOD frequency. There is another problem, however, that arises from the presence of conspecifics and other electrogenic fish, and this is where the JAR comes in.[41]

Recall that weakly electric fish sense objects around themselves as perturbations of the electric fields that are created by their EODs. If two electric fish emit EODs that are near the same frequency, the wavelike signals will constructively and destructively interfere with one another, impairing the ability of both fish to electrolocate by masking information about their motions relative to the environment. In this situation, some species of electric fish can avoid jamming by increasing the frequency difference between their own EOD and that of another fish nearby. If *Eigenmannia* detects jamming signals of a slightly lower frequency, it will raise its own EOD frequency in response, and it will lower its EOD frequency in response to slightly higher frequency jamming signals. As a result of jamming avoidance, the interference is shifted to a much higher temporal frequency where it is automatically filtered out. Other gymnotiforms have less robust jamming avoidance responses, or none at all. Some are only able to increase their own EOD frequency when confronted with a jamming signal. African mormyriforms show convergent evolution, even though they do not share an electrogenic or electroreceptive ancestor with the gymnotiforms. *Gymnarchus niloticus*, for example, has a robust JAR like that of *Eigenmannia*.[42]

The jamming avoidance response has become one of the most studied behaviors in electric fish. Its neural implementation is considerably more complicated than a relatively simple spinal reflex. In *Eigenmannia*, for example, neuronal processing for the JAR includes the coding of waveform amplitude and phase by different types of receptors on the fish's body; hindbrain processing of each stream of information independently; differential phase computation and the integration of the two information streams in the midbrain; coding of the sign of the frequency difference of the two interfering signals, and a choice of pacemaker excitation or inhibition in the diencephalon; command pulses generated in the pacemaker nucleus of the hindbrain; and electric organ discharges in the tail. At each step, Heiligenberg and his colleagues started by analyzing the computational problem that a particular neural subsystem had to solve. They created model solutions to the problem and used behavioral experiments to test their models. Then, starting with the coding of peripheral sensory neurons, they worked inward, tracing information processing pathways from one neuronal structure to the next. The choice to study the JAR has other advantages besides its intriguing neural complexity: it is social, it is easy to quantify, and unlike many other behaviors, it does not show habituation with repeated exposure to the same stimulus.[43]

Comparative neuroethological work shows not only that different gymnotiform and mormyriform species converged on the ability to generate complex electric signals but that they found different mechanisms to accomplish similar goals. There are parallels in the pacemaker cells that generate the rhythm of electric organ firing. Time coding is central to independently evolved groups, even to the point of instantiating similar computational algorithms. Both gymnotiforms and mormyriforms include species that are capable of generating complex EOD waveforms that result from firing subsets of electrocytes out of phase with one another. Different complex waveforms are associated with phylogenetic modifications of electrocyte morphology, providing evidence for the progressive evolution of increasingly complex electric organs.

Recent work with mormyrids suggests that species that evolved the ability to produce and detect subtle waveform differences diversified faster than those that did not. In addition to the JAR, patterns of electric discharge are used to threaten, signal retreat, initiate courtship, give alarm, or indicate novelty. Rays, mormyrids, and apteronotids evolved electrosensory systems independently of one another, and yet they use convergent mechanisms to adaptively cancel out electric field information resulting from their own motor movements. Physically, an electric discharge is not a propagating wave but rather an electrostatic field. As such, EODs do not echo or reverberate. Nevertheless, there are a deep similarities in time comparison and time coding between electroreception in fish, the echolocation of bats, and the auditory system of barn owls, making neuroethological comparisons a productive source of hypotheses about sensory evolution.[44]

Beyond electrosensory and electrogenic specialization, there is now a flourishing literature on fish learning and cognition. It is clear that the abilities of these animals have been systematically underestimated for some time. "Gone (or at least obsolete) is the image of fish as drudging and dim-witted pea brains, driven largely by 'instinct,' with what little behavioural flexibility they possess being severely hampered by an infamous 'three-second memory.'" Piscine evolution is and has been continually subject to niche construction effects: thousands of species of fish modify their own niches by constructing nests, spawning sites, and bowers. Hundreds of species also exhibit social learning, even those with relatively small brains. They find out from one another what is edible, where to find food, how to recognize and avoid predators, and other kinds of information crucial for their survival. Social or cognitive capabilities once thought to be the domain of primates—cultural variation, the use of social strategies, Machiavellian manipulation, cooperative hunting, and tool use—are now attributed to fish on the basis of experimental studies.

If we look deep enough into the past we find that we share many of our most basic anatomical features with fish, attributes of a distant common ancestor. The paleontologist and anatomist Neil Shubin writes that "the best road maps to human bodies lie in the bodies of other animals. The simplest way to teach students the nerves in the human head is to show them the state of affairs in sharks. The easiest road map to their limbs lies in fish." Fish have always been part of the hominin niche, too. What we are, what we've become, has been shaped by countless encounters with our distant piscine relatives over the longue durée.[45]

# Conclusion

## Nothing but a Movement of Electrons

In a sense, all life consists of the colonization of an electric world. But to see that, we have to go back to the very beginning. Within a few seconds of the big bang—an eternity in cosmological terms—the temperature of the expanding universe had dropped enough for electrons to appear. All interactions between matter seem to be governed by a few fundamental forces. Any particle that has a net electric charge, like the electron, exerts an electromagnetic force that decreases with distance according to an inverse square law and can repel as well as attract. "This force acts as the cement for most ordinary materials, including virtually everything in our homes, such as tables, chairs, books, even the kitchen sink." In the early universe, charged particles were drawn together by electromagnetic forces to form atoms, beginning with the simplest combination: one negatively charged electron and one positively charged proton, making up the element hydrogen. Helium also formed, an atom consisting of two electrons, two protons and two electrically neutral neutrons. While the electromagnetic force dominates at microscopic scales, another fundamental force prevails over longer distances: gravity. Long-range gravitational interactions drew clouds of hydrogen and helium atoms into galaxies and stars. Chemical reactions in stars produced heavier elements and provided heat and light to nearby planets.[1]

Prior to the emergence of life, billions of years later but still billions of years before the present, the gravitational field of the earth held an atmosphere in place. Its chemical composition was different than it is today, with carbon dioxide as the predominant gas. The cooling of the earth allowed the condensation of water vapor in the atmosphere, and the oceans were formed by millions of years of heavy rainfall. Atmospheric carbon dioxide gradually dissolved in the oceans. Light and heat came from the sun; lightning storms provided electrical energy; volcanic activity and natural radioactive materials provided heat. "But if (and oh! what a big if!)," Darwin wrote to Hooker in 1871, "we could conceive in some warm little pond, with all sorts of ammonia and phosphoric salts, light, heat, electricity, &c., present, that a proteine compound was chemically formed ready to undergo stillmore complex changes, at the present day such matter would be instantly devoured or absorbed, which would not have been the case before living creatures were formed." Most biologists now believe that life did emerge spontaneously from abiotic precursors, through a slow increase in molecular complexity.[2]

Chemical evolution preceded biological evolution in three stages. First, the basic raw materials of life, so-called organic molecules, must have emerged under abiotic conditions. In the 1950s, Harold Urey and Stanley Miller demonstrated that a closed retort filled with water, methane, and ammonia, after being heated and subject to electric sparking for a few days, contained several different amino acids. These are the building blocks of which the proteins of living organisms are largely composed. Subsequent work showed that variations on the experiment can produce all twenty of the basic amino acids, as well as other molecules (such as sugars) used in nucleotides, the molecular components that carry genetic information.

Second, more complex molecules must have been formed from these basic building blocks. The linkage of two amino acids requires the loss of a water molecule, and dehydration of many amino acids can result in more complex proteins. The formation of nucleic acids from sugars and nucleotide bases occurs under similar conditions. How this happened is still unknown, but heating, freezing, and catalysis are all possible explanations. Some of the larger molecules have hydrophilic and hydrophobic regions, which causes them to coil and fold in the presence of water, forming self-organized membranes. These membranes could enclose chemical solutions, isolating them from the surrounding environment.

Third, the genetic code must have originated somehow, and this is the most difficult transition to account for. There are several contending hypotheses about the emergence of stable, self-replicating molecules. One view posits the existence of a prebiotic RNA world that gave rise eventually to proteins and DNA, but there are other alternatives and the choice need not concern us here.[3]

All bulk matter, including all living matter, is held together by electromagnetic forces. Furthermore, all life is powered by electrons. When electromagnetic energy excites an electron, it drops back to a lower energy state as quickly as a hundred millionth of a second later, releasing the energy. The Hungarian biologist Albert Szent-Györgyi wrote, "Life has learned to catch the electron in the excited state, uncouple it from its partner and let it drop back to the ground-state through its biological machinery utilizing its excess energy for life's processes . . . Life is nothing but a movement of electrons!" The conversion of energy from one form to another by molecular machinery in the cell happens in very small steps. Individual photons are captured by chlorophyll molecules. Single ions move tiny quanta of electrochemical energy through semi-permeable membranes. Electrons are carried one at a time by electron transport chains. "This allows a level of control, and efficiency, that is rarely seen in our familiar macroscale world."[4]

Many different kinds of cells display a sensitivity to electric fields known as galvanotaxis. The spontaneous locomotion of these cells is biased by the direction of an external electric field, and stationary cells are stimulated by it to become motile. This occurs in single-celled organisms like amoeba and ciliates, multicellular organisms like *Volvox*, and the motile tissue cells of larger organisms (including flagellates, leukocytes, and macrophages). In the earliest organisms, communication between cells was limited to chemical diffusion. The emergence of complicated multicellular organisms required a division of labor at the cellular level and more effective communication between subsystems. It may have taken as long as a billion years for multicellular organisms to begin to evolve nerves, cells that could convert information from slow-moving chemical signals into electrical signals that could be directed through the organism much more rapidly. In organisms that were still more complex, clumps of interacting neurons formed the first nervous systems. Growing neurons themselves responded to the direction of electric fields, a phenomenon known as galvanotropism. Specialized sensory organs transduced information from light, chemical gradients, pressure waves, electric fields, and other physical stimuli into electrical signals that were processed by increasingly complex networks of neurons.[5]

Some single-celled organisms, and some motile cells of multicellular organisms, are propelled by a spinning flagellum driven by a molecular motor. In larger organisms, movement is made possible by combining great numbers of molecules capable of minute motions into tissues that make macroscopic movements. As we have seen, in a few of these organisms, muscle cells have been further modified to generate large amounts of electricity. Given the ubiquity of electricity in all cells, we might wonder why larger organisms did not evolve something resembling electric motors. The biologist and physicist Steven Vogel notes that the problem is not with rotation, since it is possible to build electric motors with linear or reciprocating movements. The probable roadblock was the lack of a good, wire-like conductor. "The salt solutions of cells don't come close. For instance, a strong solution of potassium chloride . . . an especially conductive brew, is still nine million times less conductive than copper." If nerves had conducted electricity, as people believed at one time, biological electric motors might be ubiquitous, too. Instead, "neuromuscular systems use electricity in a way peculiarly adapted to or limited by (take your choice) their low conductivity." Individual electric currents flow very short distances in the nerve, exciting another nearby current to carry the signal a tiny bit farther.[6]

All life is electric in the sense of being held together by electromagnetic forces, powered by electrons, and sensitive to electric fields and electromagnetic radiation. Animals seek actively for nutrients, in contrast to plants, and their behavior depends on electrochemical signals in sensory, nervous, and motor systems. Some animals inhabit electroreceptive worlds by sensing low frequency electric fields. Some fish are further capable of generating electric signals of more or less complexity and using them actively to sense their environment and to communicate with one another. In a few fish, electrogenesis has developed to the point where it is powerful enough to use for predation and defense. Strongly and weakly electric fish had fully colonized electric worlds by the time the hominin lineage emerged.

Unlike electric fish, human beings have not evolved mechanisms for sensing electric fields or for generating electric discharges. We do, however, possess a different kind of specialization. We readily make tools, apparatus, and equipment and ceaselessly use them to remake our own environment and those of other organisms. Here we cast our net pretty widely, as likely to use organic as inorganic materials if it suits our purposes. We manipulate ecosystems, species, individual organisms, vivisected systems, tissues, cell lines, and organic molecules, endlessly remixing and mashing them up in new configurations. And in doing so, we constantly change ourselves.

Human beings have always had the possibility of electric encounters, but they were relatively infrequent until the nineteenth century, primarily limited to experiences with lightning and strongly electric fish. Our fascination with natural sources of electricity, and our continual fiddling with them, eventually allowed us to build on these capabilities and create new ones. A *lot* of new ones. The instantaneity of electric communication changes social relationships among close kin and nation states alike, changing our possibilities for learning, amusing ourselves, conducting economic transactions, and engaging in politics. Our use of electric light isolates us from daily and seasonal variations of daylight. We interact with one another differently, live and work in different kinds of buildings and cities, control the growth of plants and animals in greenhouses and factory farms. Artificial light fills the nighttime skies over a large proportion of the inhabited earth. When electric light became widespread, it altered the circadian rhythms of many of the organisms exposed to it, not only humans or other domesticated species. There is concern now that circadian disruption may be a factor in the global increase in breast cancer. Electricity is also widely used for pumping water and sewage, for refrigeration, cooking, heating, ventilation, and air conditioning. Each of these technologies greatly increases the range of environments human beings can inhabit in relatively large numbers and changes the habitats in which we cultivate other species. We are continually confronted by the unintended consequences of generating electric power in large quantities, as we divert and dam rivers, burn coal, split atoms, and build wind and solar farms.[7]

Beyond electric power, our more recent ability to manipulate electrons with electronic devices has truly made us inhabitants of an electric world. Our senses have expanded in every direction. Radio telescopes and satellites allow us to see the cosmic microwave background radiation left over from the big bang. Atomic force microscopes let us visualize and manipulate individual atoms. Particle accelerators and detectors have revealed whole families of elementary particles that are smaller than atoms. Between the cosmological and atomic scales, transducers convert every kind of physical signal or stimulus into electric charge or current. Electronic devices to process these signals are now manufactured in almost unimaginable quantities. In 1965, the physical chemist Gordon E. Moore noticed that the number of transistors being incorporated into each new integrated circuit device was doubling at regular intervals, and he predicted that the trend would continue. More than forty years later the density of transistors per device is still doubling regularly. The number of transistors produced annually is now on the order of $10^{19}$. To put that figure in perspective, it is roughly the same as the number of individual raindrops that fall on the coterminous United States in a given year.[8]

The most notable consequence of microelectronics has been the development of digital computers to process information and the ancillary technologies that allow them to store, retrieve, and transmit it to one another. David Christian notes that, in the twentieth century, the total output of the global economy increased about twenty-fold and that "growth in just the three years from 1995 to 1998 is estimated to have been greater than total growth in the 10,000 years before 1900." In the last decade, we have witnessed an astounding acceleration in the amount of new digital information created each year. New information created in the year 2002 alone was estimated to have been on the order of 5 exabytes ($5 \times 10^{18}$ bytes). Printed as text, one megabyte takes approximately as much space as a traditional book, so storing 5 exabytes in this form would require 37,000 libraries the size of the Library of Congress. A similar study, published in 2009, estimates that Americans consumed about 3.6 zettabytes ($3.6 \times 10^{21}$ bytes) the previous year. If this information were printed as text in traditional books and they were stacked tightly side by side, they would cover the United States, including Alaska, to a depth of about seven feet.[9]

As in earlier eras, biological processes and networks have proven to be a rich source of ideas for electronic and computational systems, and those, in turn, provide us with mechanisms that appear to us to be similar to those at work in natural systems. The great advantage of digital computers is that they are behaviorally plastic: they can be programmed to simulate any process that can be given an algorithmic description. Paradigms like cellular automata, computational chemistry, neural networks, artificial immune systems, genetic algorithms, genetic programming, artificial life, agent-based simulation, swarm intelligence, subsumption architecture, and neuromorphic electronics all trade on biological metaphors to solve difficult classes of problems. At the same time, biological entities like DNA, viruses, bacteria, and tissues are now being used as computing elements in techno-organic assemblages that are as hybrid as anything dreamt up by Du Bois-Reymond.[10]

Attempting to provide any kind of overall assessment of the pluses and minuses of our move into an electric world is clearly quixotic. In analyzing electronic circuits, engineers learn to apply two conservation laws formulated in the nineteenth century by Gustav Kirchhoff. One of these says that charge is conserved: "the sum of currents into a point in a circuit equals the sum of the currents out." The other says that electrical energy is conserved: "the sum of voltage drops around any closed circuit is zero." For every positive consequence of electricity and electronics we can find a negative one, and vice versa. The health benefits of biomedical instrumentation are opposed by the risks of electrocution and exposure to e-waste and electromagnetic fields. Instant global access to information is balanced by the physical inaccessibility of the stored information. Before the twentieth century, a text could only become unreadable if the language died, but it is impossible to recover information from modern digital storage devices without an elaborate and very expensive technological infrastructure in place. In 2010, the UN International Telecommunications Union reported that the number of cell phone users in the world would soon hit 5 billion. The UN also reported that in the same year more people in India had access to a mobile telephone than to a toilet. Many handheld electronic devices follow a problematic circuit thousands of miles, from the factory where they are assembled through a brief stint in the West, only to end up months later in a vast pile of e-waste a few miles from where they began. Robotic rovers send us pictures of neighboring planets; robotic drones send pictures of war zones; robotic bombs show us a quick trip to devastation, and then nothing.[11]

Electric worlds have been discovered and inhabited by many animals before us and, assuming that life persists, will be discovered by many others once we are extinct. The kind of electric world we have made and the way we experience it are deeply human. But we are by no means its only denizens. Somewhere in the world right now, a neonatal fish is making its first tentative electric organ discharges. Whether it manages to avoid predators, find prey, live to adulthood, find a mate, and reproduce will depend on signals in an electric world. Somewhere in the world right now, electrically generated ultrasonic waves are reflecting off of a human fetus and being transduced into another electrical signal that is processed to create a computer image of the baby. Whether or not it is brought to term, has a healthy and happy childhood, avoids harm, survives to adulthood, and finds a partner will also depend on signals in an electric world. And their pasts, and perhaps their futures, are more interconnected than either can ever realize.

# Notes

### *Introduction*

1. Odling-Smee et al., "Niche Construction," quote on 641; Lewontin, "Gene, Organism, and Environment"; Odling-Smee et al., "Niche Construction"; Laland et al., "Evolutionary Consequences"; Jones et al., "Positive and Negative Effects."

2. Jones et al., "Positive and Negative Effects"; Laland et al., "Evolutionary Consequences."

3. Smith, "Ultimate Ecosystem Engineers"; Smail, *Deep History*; E. Russell, *Evolutionary History*; Lehmann, "Adaptive Dynamics," quote on 560.

### *Chapter One. Strongly Electric Fish*

1. Moller, *Electric Fishes*, 61–76, 443–46; "*Malapterurus electricus* (Gmelin, 1789)," Fishbase.org. Zoogeography: Golubtsov & Berendzen, "Morphological Evidence." Feeding: Belbenoit et al., "Ethological Observations"; Sagua, "Observations." Form: Alexander, "Structure and Function"; Lissmann, "On the Function," "sausage" quote on 180. Species distribution: Golubtsov & Berendzen, "Morphological Evidence."

2. Biogeography: Briggs, "Biogeography"; Stewart, "Freshwater Fish"; Otero et al., "First Description"; Leakey et al., "Lothagam"; Moller, *Electric Fishes*. Electric shocks: Alves-Gomes, "Evolution"; Sullivan et al., "Phylogenetic"; Lundberg et al., "Discovery." Drinking water: Erlandson, "Aquatic Adaptations."

3. Tattersall, *World from Beginnings*, "slog" quote on 41; Christian, *Maps of Time*. Adaptive radiation: Andrews, "Evolution and Environment." Climate: "Miocene" in Hancock & Skinner, *Oxford Companion*. Extinction and speciation: Vrba, "Ecological and Adaptive Changes."

4. Tattersall, *World from Beginnings*; Lewin, "Origin"; Christian, *Maps of Time*, chap. 6. Walking efficiency: Pontzer et al., "Metabolic Cost."

5. R. L. Lyman, "Archaeofaunas"; Plummer, "Flaked Stones"; Sept, "Archaeological Evidence"; Laland et al., "Niche Construction"; Tattersall, *World from Beginnings*. Focus on mammals: Erlandson, "Aquatic Adaptations"; Steele, "Unique Hominin Menu."

6. Stewart, "Early Hominid Utilisation"; Wrangham et al., "Shallow-Water Habitats."

7. Stewart, "Early Hominid Utilisation."

8. Steele, "Unique Hominin Menu"; Braun et al., "Early Hominin Diet."

9. Animal tools: Ambrose, "Paleolithic Technology"; Boesch, "Is Culture"; Laland & Hoppitt, "Do Animals Have Culture?"

10. Tattersall, *World from Beginnings*; Sept, "Archaeological Evidence"; Plummer, "Flaked Stones."

11. Hill, "Disarticulation"; Hill & Behrensmeyer, "Disarticulation Patterns"; Lyman, "Archaeofaunas."

12. Sept, "Archaeological Evidence"; Plummer, "Flaked Stones," quote on 118.

13. Tattersall, *World from Beginnings*. Tools suited to tasks: Walker, "Butchering."

14. Mithen, "Evolution of Imagination"; Davidson & McGrew, "Stone Tools"; S. Kuhn, "Evolutionary Perspectives." Novices: Geribàs et al., "Novice Knappers."

15. Chimpanzees: Boesch & Boesch, "Mental Maps"; Boesch, "Culture"; Mercader et al., "Excavation"; Davidson & McGrew, "Stone Tools."

16. Davidson & McGrew, "Stone Tools," quote on 803; Plummer, "Flaked Stones"; Mithen, "Evolution of Imagination"; Kuhn, "Evolutionary Perspectives."

17. Mercader et al., "Excavation"; Byrne, "Manual Skills"; Davidson & McGrew, "Stone Tools." Affordances: Gibson, "Theory of Affordances."

18. Mithen, "Evolution of Imagination"; Davidson & McGrew, "Stone Tools," quote on 811; Whiten et al., "Evolution and Cultural Transmission." World as extension of mind: Clark, *Being There*; Clark, *Natural-Born Cyborgs*; Sterelny, *Thought in a Hostile World*.

19. Arthur, *Nature of Technology*, quote on 170.

20. Tattersall, *World from Beginnings*; Joordens et al., "Relevance"; "*Clarias batrachus* (Linnaeus, 1758)," Fishbase.org; Erlandson, "Aquatic Adaptations"; Erlandson & Moss, "Shellfish Feeders."

21. Tattersall, *World from Beginnings*; Steele, "Unique Hominin Menu." Ethiopia: Trapani, "Quaternary Fossil Fish"; Fleagle et al., "Paleoanthropology"; Stewart & Murray, "Fish Remains."

22. South Africa: Andel, "Late Pleistocene"; Klein et al., "Yserfontein 1";

Marean, "Sea." Refugia and corridors: Bailey et al., "Coastal Shelf"; Fa, "Tidal Amplitude." Human population crash: Marean, "Sea"; Marean et al., "Early Human Use."

23. Klein et al., "Yserfontein 1," quote on 5708; McBrearty & Brooks, "Revolution"; Habgood & Franklin, "Revolution"; Steele, "Unique Hominin Menu."

24. Erlandson, "Aquatic Adaptations"; Erlandson & Moss, "Shellfish Feeders"; Fa, "Tidal Amplitude."

25. Erlandson, "Aquatic Adaptations"; Erlandson & Rick, "Archaeology Meets Marine Ecology." Meighan and Washburn & Lancaster quoted in Erlandson, "Aquatic Adaptations," on 287, 288. Gunditjmara: Builth, "Gunditjmara." Hominin colonization of Australia: Habgood & Franklin, "Revolution." Homo erectus: Morwood et al., "Fission-Track Ages."

26. Erlandson, "Aquatic Adaptations," table 1 on 306–9; McBrearty & Brooks, "Revolution."

27. Moller, *Electric Fishes*; "*Torpedo torpedo* (Linnaeus, 1758)," "*Torpedo marmorata* (Risso, 1810)" and "*Torpedo nobiliana* (Bonaparte, 1835)," Fishbase.org; Serena, *Field Identification Guide*; Fischer et al., "Batoid Fishes." Waders: R. Aidan Martin, "Biological Batteries," *Diver Magazine* (April 1995).

28. Moller, *Electric Fishes*; "*Narcine tasmaniensis* (Richardson, 1841)," "*Hypnos monopterygius* (Shaw, 1795)," and "*Torpedo californica* (Ayres, 1855)," Fishbase.org; Carvalho et al., "Order Torpediniformes"; Commonwealth Bureau, *Official Year Book*, 757.

29. Erlandson, "Aquatic Adaptations"; Erlandson & Rick, "Archaeology Meets Marine Ecology"; Erlandson & Moss, "Shellfish Feeders."

30. Colonization by biomes: Dixon, "Human Colonization." *Torpedo californica*: California, "Skates" (2010), quote on 5–10.

31. Erlandson & Moss, "Shellfish Feeders"; Erlandson & Rick, "Archaeology Meets Marine Ecology"; Erlandson, "Aquatic Adaptations"; Bray & Hixon, "Night-Shocker"; Eschmeyer & Herald, *Field Guide*, 53–54; California, "Skates" (2001, 2010).

32. Moller, *Electric Fishes*; "*Electrophorus electricus* (Linnaeus, 1766)," Fish base.org; Westby, "Ecology"; Coates, "Kick," quote on 76; Crampton, "Gymnotiform Fish."

33. Africa: Murray, "Supplemental Observations," "unlettered savage" quote on 114; Baird, *Cyclopaedia*, 342–343; Hutchinson, "Social and Domestic Traits"; Chapin, "Date: 12/28/1909 to 1/10/1910," 16; Weeks, "Anthropological Notes"; Henry & Colman, *Soul*, 136, 297. South America: Markham, "New Discovery"; Asúa, "Experiments."

34. Kpelle: Westerman & Schütze, *Kpelle*, 161. Mende: Migeod, *Mende*

*Natural History Vocabulary*, 24. Hausa: Robinson, *Dictionary*, 166; Newman, *Hausa-English Dictionary*, 155. Bobangi: Whitehead, *Grammar*, 204, 208. Arabic: Wehr, *Dictionary*, 345; Wilson, "On the Electric Fishes" says root is shared with "trembling," etc., but not "thunder." Persian: Wollaston, *English-Persian Dictionary*, 9, 114, 117, 280, 380. Greek: Thompson, "On Egyptian Fish-Names"; Kellaway, "Part Played"; Moller, *Electric Fish*, 59. Latin: Cicero, *De natura deorum* 2.50.127; Lewis & Short, *Latin Dictionary*. Brazilian Portuguese: Chamberlain & Harmon, *Dictionary*, quote on 498. Spanish: Velazquez de la Cadena, *Dictionary*, 403. General information about languages of the world: Ethnologue.com.

35. Stewart, "Early Hominid Utilisation"; Stewart, "Fossil Fish." Narmer palette: el-Shahawy, "Narmer Palette"; Wengrow, "Narmer"; Fairservis, "Revised View"; T. Wilkinson, "What a King Is This."

36. Bronk Ramsey et al., "Radiocarbon-Based Chronology"; Malek, "Old Kingdom"; J. G. Wilkinson, *Manners and Customs*, xxii–xxiii, 21–22; Horapollo, *Hieroglyphics*, 95. There is a detailed online guide to the Mastaba of Ty at www.osirisnet.net/mastabas/ty/e_ty_01.htm.

37. Archestratus: Radcliffe, *Fishing*, 161 n.1. Torpedo as food: Dalby, *Food*; van der Eijk, "Role"; Pliny, *Natural History*, 9.67; Apicus, *Cookery*, 211. Pompeii: Anonymous, "More Marine Stories"; Drewer, "Fishermen"; Wu, "Electric Fish."

38. Therapeutic uses: Pliny, *Natural History*, 32.47 (depilatory), 32.46 (parturition), 32.50 (anaphrodisiac). The anaphrodisiac prescription was repeated as late as Jacques Ferrand's *Treatise on Lovesickness* (1610). Kellaway, "Part Played"; Copenhaver, "Tale."

39. Pliny, *Natural History*, 32.33; Kellaway, "Part Played," quote on 130; Copenhaver, "Tale"; Lovell, *Sive Panzoologicomineralogia*, quote on 191.

40. Lloyd, "Invention of Nature," quote on 418; Grant, *History of Natural Philosophy*, quote on 1.

41. Aristotle, *History of Animals*, 37, 41, 104, 109, 151, "stupefies" quote on 254–55; Radcliffe, *Fishing*, 110–15; Gudger, "Five Great Naturalists."

42. Kellaway, "Part Played," Plutarch quote on 116; Copenhaver, "Tale"; Debru, "Power"; Denkinger, "Arcadia"; Sharples, *Theophrastus*, 98–101.

43. Hero, *Pneumatics*, 9–10. Strato: Lindberg, *Beginnings*, 74–75.

44. Pliny, *Natural History*, 32.2; Copenhaver, "Tale," "pneumata" quote on 379. Kaempfer: from Pliny 32.2 n. 2 in edition cited; see also translation in Carrubba & Bowers, "First Report."

45. Copenhaver, "Tale"; Debru, "Power"; Chalmers, "Lodestone"; Piccolino, "Taming."

46. Dunn, "Galen"; Reeves & Taylor, "History"; Thorndyke, "Mediaeval Magic"; Caputi, "Contributions"; Home, "Electricity"; Chalmers, "Lodestone."

47. Guerrini, *Experimenting*; Lloyd, "Transformations"; Reeves & Taylor, "History."

48. Guerrini, *Experimenting*. Fish mummies: Brier, "Autopsies."

49. Lloyd, "Transformations."

*Chapter Two. Modeling Animal Electricity*

1. Benyus, *Biomimicry*.

2. *30,000 Years of Art*, 4 (Lion Man), 6 (horses), 11 (chimera), 51 (lion-headed female), 100 (Burney relief).

3. Cotterell & Kamminga, *Mechanics*.

4. Keyser, "Purpose"; Kanani, "Parthian"; Stillings, "Mediterranean Origins." König also hypothesized that the Parthian cells might be used to disinfect water.

5. Edgerton, *Shock of the Old*; Keyser, "Purpose," quote on 98.

6. Byzantine: Aegineta, *Medical Works*, 48, 334, 355; Aegineta, *Seven Books*, 266, 570; Schechter, "Origins, Part 1," "Origins, Part 2"; Stillings, "Piscean Origin"; Stillings, "Mediterranean Origins"; Abd al-Latif, *Relation*, 145–46; Lanza, "Electric Fish," Abd al-Latif quote on 27; Thompson, "Egyptian Fish Names." Albert the Great is quoted in Copenhaver, "Tale," 383.

7. Ben-Amos, "Men and Animals," quote on 245; Blier, *Royal Arts*; Moller, *Electric Fish*, 11–12. Armlet: Pitt Rivers Museum, Oxford, UK, 1991.13.26.

8. Copenhaver, "Renaissance"; Gudger, "Five Great Naturalists." Plato, *Meno*. Erasmus: *Adages*, 56, 89. Paresthesia: Halstead, "Poisonous Fishes."

9. Grafton et al., *New Worlds*; Myers, "Brief Sketch"; Léry, *History*, 96, 114; Delbourgo, *Most Amazing*.

10. Copenhaver, "Tale"; Carrubba & Bowers, "First Report," quoted translation of Kaempfer on 272, 273.

11. Purchas, *Haklytus Posthumus*, 9:227–28, 287–88; Koehler et al., "Eels"; Lanza, "Electric Fish," Godinho quote on 27; Moller, *Electric Fishes*, xxii, 12.

12. Asúa, "Experiments"; Koehler et al., "Eels," Richer quote on 724; Purchas, *Haklytus Posthumus*, 16:488 (Cardim); Olmstead, "Scientific Expedition"; Janet Todd, *Secret Life*, 57; Behn, *Oroonoko*, 51–52; Evelyn, *Diary*; for Europeans eating exotic animals, see Ritvo, *Platypus*, chap. 5.

13. Montaigne, "Apology"; Copenhaver, "Tale," quote on 384; Copenhaver, "Renaissance"; Heilbron, *Electricity*, chap. 1; Gudger, "Five Great Naturalists."

14. Copenhaver, "Tale"; Thorndyke, "Mediaeval Magic"; Dear, "*Totius in Verba*." The "stupefactive emanation" quote comes from Walter Charleton.

15. Piccolino, "Taming"; Whitteridge, "Physics and Chemistry"; Asúa, "Ex-

periments"; Koehler et al., "Eels"; Wu, "Electric Fish"; Lanza, "Electric Fish"; Moller, *Electric Fish*, 14; Heilbron, *Electricity*, 198–202; Daniell, "Animal Electricity," Lorenzini quoted on 497.

16. Lanza, "Electric Fish"; Moller, *Electric Fish*, 14; Brazier, *17th & 18th Centuries*; Whitteridge, "Physiology."

17. Cipolla, *Clocks*; Crosby, *Measure*.

18. Gilbert, *Lodestone*; Chalmers, "Lodestone"; Hackmann, "Relationship"; Heilbron, *Electricity*, chap. 3; Brazier, *17th & 18th Centuries*, 6–10.

19. Heilbron, *Electricity*, chaps. 3, 6, 8, 13; Brazier, *17th & 18th Centuries*, 10–11; Keithley, *Story*, 9–13, 15; Fara, *Entertainment*, chaps. 2, 3; Hackmann, "Relationship"; Hackmann, *Electricity*; Rossi, *Birth*, 148–49; Schaffer, "Natural Philosophy."

20. Heilbron, *Electricity*, chap. 8; Fara, *Entertainment*, chap. 3; Brazier, *17th & 18th Centuries*, 176–78; Elsenaar & Scha, "Electric Body Manipulation."

21. Heilbron, *Electricity*, chaps. 8, 9; Fara, *Entertainment*, chap. 3.

22. Heilbron, *Electricity*, chap. 13, quote on 311; Fara, *Entertainment*, chap. 3; Riskin, *Science*, 88.

23. Heilbron, *Electricity*, chap. 13, Musschenbroek quoted on 313.

24. Ibid., quote on 317.

25. Piccolino, "Taming"; Koehler et al., "Eels"; Wu, "Electric Fish"; Millingen, *Curiosities*, 401.

26. Ingram, "New Experiments," quotes on 50, 52; Koehler et al., "Eels."

27. Turner, *Electricology*, 28, 29.

28. Piccolino, "Taming," Adanson quote on 128; Riskin, *Science*; Moller, *Electric Fish*, 15.

29. Koehler et al., "Eels." Immutable mobiles: Latour, *Science in Action*, 229.

30. Koehler et al., "Eels"; 's Gravesande quote on 729; Finger, "Dr. Alexander Garden."

31. Koehler et al., "Eels."

32. Ibid., Fouchy quote on 745.

33. Bancroft, *Essay*, 190–200; quotes on 192, 194, 196, 198; Finger, "Edward Bancroft's 'Torporific Eels.'"

34. John Walsh to Benjamin Franklin, July 12, 1772, published in Franklin, *Works*, 6:348–50; Jungnickel & McCormmach, *Cavendish*, 187–89; Ingenhousz, "Extract," "elbow" quote on 2; Schaffer, "Fish and Ships" (Schaffer incorrectly uses the term *stingray* to refer to electric rays, in this paper and in "Exactly Like a Stingray"). Bodies in circuits: Schaffer, "Self Evidence."

35. John Walsh to Benjamin Franklin, July 12, 1772, published in Franklin, *Works*, 6:348–50; Walsh, "Electric Property," quotes on 465, 468; Piccolino &

Bresadola, "Drawing a Spark"; Daniell, "Animal Electricity," Walsh "animate phials" quote on 502.

36. Piccolino & Bresadola, "Drawing a Spark," 57; Moller, *Electric Fish*, 40; Jungnickel & McCormmach, *Cavendish*, 189; Piccolino & Bresadola, "Drawing a Spark." Electrometers: Priestley, "Account."

37. Cavendish, "Account," quote on 198; Jungnickel & McCormmach, *Cavendish*, 189; Heilbron, *Electricity*, chap. 19; Berry, "Aspects."

38. Hunter, "Anatomical Observations"; Cavendish, "Account," quote on 200; Jungnickel & McCormmach, *Cavendish*, 187–90. Battery of Leyden jars: Benjamin Franklin to Thomas Hubbard, "Electrical Apparatus—Description of a Battery" (28 April 1758), in Franklin, *Works*, 5:361–63.

39. Cavendish, "Account"; Jungnickel & McCormmach, *Cavendish*, 189–90.

40. Cavendish, "Account," "considerable difference" quote on 212; Jungnickel & McCormmach, *Cavendish*, 186–87; Heilbron, *Electricity*, chap. 19.

41. Dear, "*Totius in verba*," quote on 152; Schaffer, "Self-Evidence," Boyle quote on 328. See also Dear, "From Truth to Disinterestedness"; Daston, "Objectivity."

42. Cavendish, "Account"; Ronayne, "Letter"; Maxwell in Cavendish & Maxwell, *Electrical Researches*, xxxvii–xxxviii, Ronayne quote from xxxvii; Jungnickel & McCormmach, *Cavendish*, 189–90.

43. Jungnickel & McCormmach, *Cavendish*, 189–90; Heilbron, *Electricity*, chap. 19; Schaffer, "Fish and Ships," quote on 80. Thunder houses: Schaffer, "Fish and Ships," quote on 77; Delbourgo, *Most Amazing*.

### *Chapter Three. Electrophysiology*

1. Brazier, *17th & 18th Centuries*, chaps. 3, 4; Verkhratsky et al., "From Galvani"; Smith, "Brain and Mind."

2. Brazier, *17th & 18th Centuries*, chap. 4; Brazier, "Historical Development"; Whitteridge, "Physiology"; Haller, "Dissertation," quotes from 657–62.

3. Newton: Newton, *Principia*, 547; Hall & Hall, "Introduction"; Home, "Newton"; Home, "Electricity"; W. Wallace, "Vibrating Nerve Impulse"; Yolton, "Physiology"; Borelli & Croone: Brazier, *17th & 18th Centuries*, chaps. 4, 13; Glassman & Buckingham, "David Hartley's Neural Vibrations."

4. Haller, *First Lines*, "ligature" quote from 1:221; Haller, "Dissertation"; Home, "Electricity"; Brazier, "Historical Development," "substantial" quote on 14.

5. Brazier, "Historical Development," Whytt quote on 33; Yolton, "Physiology"; Rocca, "William Cullen."

6. Brazier, "Historical Development"; Haller, "Dissertation," "irritability" quote on 691; Koehler, "Neuroscience"; La Mettrie, *Man a Machine*, "machine" quote on 71–72, "springs" and "feelings" quotes on 62; Otis, *Networking*.

7. Glassman & Buckingham, "David Hartley's Neural Vibrations"; W. Wallace, "Vibrating Nerve Impulse"; *Stanford Encyclopedia of Philosophy*, "David Hartley" at http://plato.stanford.edu/entries/hartley/; Home, "Electricity."

8. Guerrini, "Alexander Monro *Primus*," quotes on 1, 7. Scientific production as performance: Schaffer, "Natural Philosophy."

9. Priestley, *History*, 104, 173–74; Elsenaar & Scha, "Electric Body Manipulation."

10. Guerrini, *Experimenting*; Guerrini, "Alexander Monro *Primus*." Cudgel-playing: Ash, *New and Complete Dictionary of the English Language* (1775), s.v. "cudgelplayer," "cudgelplaying," http://books.google.com/books?id=LDNAAAAAYAAJ.

11. Schaffer, "Self-evidence," quotes on 333–34; Bertucci, "Sparks in the Dark"; Roberts, "Science"; Schaffer, "Natural Philosophy"; Collins, *Changing Order*; Elsenaar & Scha, "Electric Body Manipulation"; Heilbron, "Franklin."

12. *Gentleman's Magazine* 15 (1745): 193–97. Early issues of the *Gentleman's Magazine* are online at the Hathi Trust digital library at http://catalog.hathitrust.org/Record/000542092. Heilbron, "Franklin," quotes on 539, 541; Locke & Finger, "*Gentleman's Magazine*."

13. *Gentleman's Magazine* 16 (1746): 163 (quote), 291–92, 356. Locke & Finger, "*Gentleman's Magazine*"; Finger & Ferguson, "Role."

14. Bertucci, "Therapeutic Attractions," Nollet quoted on 273; Locke & Finger, "*Gentleman's Magazine*"; Bertucci, "Shocking Bag"; Finger, "Luigi Galvani"; Priestley, *History*, 167.

15. Bertucci, "Therapeutic Attractions"; Locke & Finger, "*Gentleman's Magazine*"; Schaffer, "Natural Philosophy"; Elsenaar & Scha, "Electric Body Manipulation"; Priestley, *History*, "perfumed" on 180.

16. Bertucci, "Therapeutic Attractions"; Priestley, *History*, 179–91; Schaffer, "Natural Philosophy," quote on 13.

17. Bertucci, "Therapeutic Attractions," 278; Bertucci, "Shocking Bag"; Mauro, "Role"; Beard, *Practical Treatise*, Electro-Therapeutics chap. 1, 200; Wesley, *Desideratum*, 42–72; Schiller, "Reverend Wesley."

18. Koehler et al., "Eels."

19. Ibid., Van der Lott quoted on 741; Piccolino, *Taming of the Ray*, 35–36; Haüy, "Electricity." Slave demonstrations: Delbourgo, *Most Amazing Scene*, 186; Bryant, "Account."

20. Delbourgo, *Most Amazing Scene*, quote on 187. Kaempfer: Carrubba & Bowers, "First Report," quote on 273. Termeyer and Eder: Asúa, "Experiments." Flagg, "Observations," quote on 171.

21. "Historical Account," *Gentleman's Magazine* 15 (1745), quote on 194;

Bertucci, "Sparks in the Dark"; Fara, "Attractive Therapy"; Plumb, "Electric Stroke," "dullest" quote on 87; Harvey, *Reading Sex*, 87 (quote), 107, 124–34.

22. Focaccia & Simili, "Luigi Galvani"; Bresadola, "At Play with Nature," Galvani quote on 70; Bresadola, "Medicine and Science"; Piccolino, "Luigi Galvani"; Piccolino, "Animal Electricity."

23. Galvani quotes from translated excerpt of *De viribus electricitatus* in Clarke & O'Malley, *Human Brain and Spinal Cord*, 180. Piccolino, "Luigi Galvani." Wife and servants: Bertucci, "Marco Piccolino"; Parent, "Giovanni Aldini."

24. Bresadola, "Medicine and Science," 376; Piccolino, "Luigi Galvani"; Piccolino, "Animal Electricity"; Mauro, "Role," Volta quoted on 144; Otis, "Metaphoric Circuit"; Sanford, "Contact Electrification"; Kipnis, "Luigi Galvani." *De viribus electricitatus* was dated 1791 but actually published in early 1792.

25. Piccolino, "Luigi Galvani"; Pancaldi, *Volta*; Lanza, "Experiment," Galvani quote on 596; Adams, "On Grove's."

26. Piccolino, "Animal Electricity," quotes on 387, 388; Pancaldi, *Volta*, 179–80; Gill, "Voltaic Enigma"; Mauro, "Role."

27. Canton, "Letter"; Cavallo, "Methods"; Elliott, "Abraham Bennet"; Nicholson, "Description"; Donaldson, "Mr. Nicholson's Doubler"; Sanford, "Discovery"; Kipnis, "Scientific Controversies."

28. Nicholson, "Observations," quote on 358; Schaffer, "Fish and Ships," 86–87.

29. Pancaldi, *Volta*, quote on 4; Volta, "On the Electricity," "*organe*" quote on 405; Kipnis, "Scientific Controversies"; Mauro, "Role," "active" quote on 147; Gill, "Voltaic Enigma."

30. 30. Mertens, "Shocks and Sparks," Volta quote on 302.

31. Ibid., Volta quotes on 302–3; Sudduth, "Voltaic Pile"; Pancaldi, "Hybrid Objects."

32. Richards, *Romantic Conception*, 317–21; Humboldt's "light" quote on 321; Agazzi, "Impact," Humboldt's "very strange" quote on 43; Kipnis, "Luigi Galvani"; Kettenmann, "Alexander von Humboldt," Mauro, "Role"; Brazier, *19th Century*, chap. 2.

33. Humboldt, *Personal Narrative* 4, quotes on 343, 344.

34. Ibid., quotes on 344, 347, 348, 350, 352.

35. Ibid., quotes on 357, 359–60.

36. Humboldt and Gay-Lussac, "Experiments," quotes on 356, 357.

37. Humboldt, *Personal Narrative* 2, quote on 186; Richards, *Romantic Conception*, quotes on 129, 134, Schelling "system of nature" quote on 133; Kleinert, "Volta"; Martins, "Ørsted," Schelling "simplicity" quote on 361; Otis, *Müller's Lab*, chap. 1.

38. Darnton, *Mesmerism*, quotes on 10–11; Martins, "Ørsted."

39. Pattie, "Mesmer's Medical Dissertation," "gravity" quote on 278; Lanska & Lanska, "Franz Anton Mesmer," Mesmer quoted on 302; Smith, "Brain and Mind"; Parent, "Giovanni Aldini."

40. Lanska & Lanska, "Franz Anton Mesmer"; Pattie, "Mesmer's Medical Dissertation," "features" quote on 282.

41. Lanska & Lanska, "Franz Anton Mesmer."

42. Ibid.; Schaffer, "Self Evidence."

43. Lanska & Lanska, "Franz Anton Mesmer," quote on 307; Darnton, *Mesmerism*; Franklin, *Animal Magnetism*, quote on 13; Fara, "Attractive Therapy"; Schaffer, "Self Evidence"; *Encyclopedia of Occultism & Parapsychology*, 5th ed., s.v. "Mesmer, mesmerism."

*Chapter Four. The Spark of Life*

1. Schechter, "Early Experiences," quote from the *Registers of the Royal Humane Society of London* on 362.

2. Driscol et al., "Remarkable Dr. Abildgaard," quotes on 878, 879.

3. Schechter, "Early Experiences," Humboldt quoted on 361; Fowler, *Experiments*; Althaus, *Treatise*, 156; Garratt, *Medical Electricity*, 250.

4. Althaus, *Treatise*, 155–60; Haüy, "Electricity"; Fowler, *Experiments*, quotes on 85, 87.

5. Nysten, *Nouvelles expériences galvaniques*; Elsenaar & Scha, "Electric Body Manipulation," "axe" quote on 23; Lunn, "Electricity," 125; Hands, "Animal Electro-Magnetism"; Althaus, *Treatise*, 293–94; Haüy, "Electricity."

6. Boling et al., "Historical Contributions"; Aldini, *Essai*; Brazier, "Historical Development"; Sleigh, "Life"; Parent, "Giovanni Aldini," Aldini quoted on 580 (Parent's translation).

7. Aldini, *Essai*, "human machine" on 141; Parent, "Giovanni Aldini"; Brazier, *19th Century*, chap. 1.

8. Aldini, *Essai*; Parent, "Giovanni Aldini," Aldini's "so much increased" quote is on 581, taken from the appendix to an English translation of his work; George Foster's trial: *Old Bailey Proceedings Online* (www.oldbaileyonline.org, v7.0, April 2012), January 1803 (t18030112-86); *Newgate Calendar*, "George Foster," quote on 318; Sleigh, "Life"; Morus, "Galvanic Cultures"; Morus, "Radicals"; Morus, *Frankenstein's Children*, chap. 5. Note that Aldini says the experiments with Foster's body took place on January 17, but the *Newgate Calendar* indicates that it was January 18. Foster's surname is given as "Forster" in some sources, and his first name variously as "John" and "Thomas."

9. *The Times* has an online archive (1785–1985), http://archive.timesonline.co.uk. Robertson: Marion, *Wonders*, 188–91.

10. Surr, *Winter*, 2:179; satirical poem reprinted in Schechter, "Early Experiences," 362.

11. Sleigh, "Life," Aldini quote on 240, Sleigh's emphasis; Mottelay, *Bibliographical History*, 305; Sachs, *Corpse*; Brown-Sequard, "Croonian Lecture."

12. Ure, "Galvanism," quote on 480; Morus, "Galvanic Cultures"; Morus, "Grand and Universal"; Morus, *Frankenstein's Children*, chap. 5.

13. Shelley, "Original Correspondence," "I saw" quote on 682; Mellor, "Making a 'Monster,'" Shelley's "dream" quote on 10.

14. Shelley, *Frankenstein*, quote on 90; Finger & Law, "Karl August Weinhold," Weinhold quote from 169; Mellor, "Making a 'Monster,'" quote on 10.

15. Clark, *Natural-Born Cyborgs*, quotes on 7.

16. Hankins, *Science*, chap. 4; Sudduth, "Eighteenth-Century Identifications"; Strickland, "Galvanic Disciplines."

17. Jungnickel & McCormmach, *Cavendish*, 147 (quote).

18. Ibid., 191, 264–77; Berry, "Some Aspects"; Hankins, *Science*, chap. 4.

19. Sudduth, "Eighteenth-Century Identifications"; Mosini, "Chemistry"; Fulhame, *Essay*, chap. 10; Mills, "Early Voltaic Batteries"; Pancaldi, "Hybrid Objects"; Sudduth, "Voltaic Pile," "throw light" from the *Morning Chronicle*, May 30, 1800, quoted on 27.

20. Sudduth, "Voltaic Pile"; Mosini, "Chemistry"; Bernardi, "Controversy"; Berry, "Some Aspects"; Strickland, "Galvanic Disciplines"; Clarke & Jacyna, *Nineteenth-Century Origins*, chap. 5; Golinski, *Science*, chap. 7.

21. Knight, *Humphry Davy*, quote on 24; Golinski, *Science*, chap. 7.

22. Knight, *Humphry Davy*; Mills, "Early Voltaic Batteries"; Pancaldi, "Hybrid Objects."

23. Knight, *Humphry Davy*, Knight's "inanimate" quote on 58, Davy's "immediately" quote on 64; Knight, "Physical Sciences."

24. Knight, *Humphry Davy*; Golinski, *Science*, Davy quote on 200; Mills, "Early Voltaic Batteries," power estimate on 396; Mottelay, *Bibliographical History*, 340–44.

25. Pancaldi, "Hybrid Objects"; Knight, *Humphry Davy*; Humphry Davy, "Account," "instrument" quote on 16.

26. Martins, "Ørsted"; Agazzi, "Impact"; Friedman, "Kant"; Knight, "Physical Sciences."

27. Adams, "On Grove's"; Gill, "Voltaic Enigma."

28. Victims of lightning injury sometimes exhibit a temporary fernlike marking of the skin that is also known as a Lichtenberg figure. Ritter may have been aware of this phenomenon, too, when he visualized the polarity of the nervous system.

29. Ritter, *Key Texts*, quotes on 275, 277, 279, 333; Hankins & Silverman, *Instruments*, Ritter's "book of nature" and "hieroglyph" quotes on 132; Clarke & Jacyna, *Nineteenth-Century Origins*, chap. 5; Schaffer, "Self-Evidence"; Christensen, "Ørsted's Concept"; Brazier, *19th Century*, chap. 2.

30. Clarke & Jacyna, *Nineteenth-Century Origins*, chap. 5, Treviranus quoted on 186; W. Wollaston, "Agency," quote on 488.

31. John Todd, "Some Observations," quote on 125; idem., "Account."

32. Oersted, "Magnetism," quotes on 118, 119; Stauffer, "Persistent Errors"; Stauffer, "Speculation"; Martins, "Ørsted"; Friedman, "Kant"; Purrington, "Electromagnetism."

33. Purrington, "Electromagnetism"; Hofmann, *Ampère*; Darrigol, "Foundations."

34. Bichat, *Physiological Researches*, quote on 1; Clarke & Jacyna, *Nineteenth-Century Origins*, chap. 7; Brazier, *17th & 18th Centuries*, chap. 12, Guerrini, *Experimenting*, chap. 4.

35. Guerrini, *Experimenting*, chap. 4; Ritvo, "Toward a More Peaceable Kingdom"; Brazier, *19th Century*, chap. 5; Fitzsimons, "Physiology"; Hoff & Geddes, "Rheotome," Nobili quote on 213; *London Encyclopedia*, vol. 8, s.v. "Electro-galvanism." The language of the nineteenth-century *Encyclopedia* is gendered, of course. I retained the pronoun "his" because I suspect that in this case both writer and readers would have expected the experimenter to be male.

36. Maceroni, "Account," quotes on 94–95; Morus, "Galvanic Cultures"; Morus, *Frankenstein's Children*, chap. 5.

37. Knellwolf & Goodall, introduction; Pocock, "Andrew Crosse."

38. Crosse, "On the Production," quote on 242; Pocock, "Andrew Crosse"; Secord, "Extraordinary Experiment."

39. Crosse, "On the Production," quotes on 243, 244; "Extraordinary Experiment," *Morning Chronicle*, January 5, 1837, "German naturalist" quote; Secord, "Extraordinary Experiment," "creation" quote on 347; Pocock, "Andrew Crosse"; Klotz & Katz, "Two Extraordinary Electrical Experiments"; [C. Crosse], *Memorials*.

40. Pocock, "Andrew Crosse," farmer quoted on 194; Secord, "Extraordinary Experiment," Faraday quotes and Secord's interpretation on 351; Hirshfeld, *Electric Life*; Melton, *Encyclopedia of Occultism & Parapsychology*, s.v. "Crosse, Andrew."

41. C. Wilkinson, *Elements*, chap. 35, quotes on 436, 439.

42. Ibid., quotes on 463, 467; Stevenson, "Suspended Animation," "anesthetists" quote on 482.

43. Page, "Henry Hill Hickman," Hickman quoted on 37; Stevenson, "Suspended Animation," Hickman quoted on 486.

44. Stevenson, "Suspended Animation"; Simpson, "Local Anaesthesia," quotes on 366, 367, 369.

45. "Traité des signes"; Richardson, "Lettsomian Lectures"; Richardson, "Croonian Lecture"; Behlmer, "Grave Doubts"; Madea & Henssge, "Timing."

### *Chapter Five. Evolutionary Theories*

1. E. Darwin, *Zoonomia*, quotes on 10, 498; Browne, *Charles Darwin Voyaging*, chap. 2; Gardner-Thorpe & Pearn, "Erasmus Darwin"; Smith, "Brain and Mind"; Pancaldi, "On Hybrid Objects."

2. Appel, "Geoffroy," Geoffroy quoted on 78, 79; Gillispie, "Scientific Aspects"; Haüy, "Electricity."

3. Desmond, *Politics*, chaps. 2–3; Paley, *Natural Theology*, quotes on 372; E. S. Russell, *Form and Function.*

4. C. Wilkinson, *Elements*; Geoffroy, "Sur l'anatomie"; Appel, "Geoffroy"; Humboldt, "Experiments"; Desmond, *Politics*, chap. 2.

5. E. S. Russell, *Form and Function*, Oken quoted on 90; Desmond, *Politics*; Gould, *Ontogeny*; Lenoir, *Strategy*, chap. 2, Meckel quote on 60; Northcutt, "Evolution"; Ritter, *Key Texts*, 135, 137.

6. Lenoir, *Strategy*, quote on 87; Desmond, *Politics*; E. S. Russell, *Form and Function.*

7. Chambers, *Vestiges*, "Almighty Deviser" quote on 168; Secord, introduction, authorship quotes on xl, xli–xlii, "entertaining mix" quote on xx–xxi, "forget" quote on xliii; Secord, *Victorian Sensation*; Desmond, *Politics*. Circulation of *Vestiges*: Secord, *Victorian Sensation.*

8. Chambers, *Vestiges*, "Almighty Deviser" quote on 168, Babbage quoted on 210, "unquestionably" quote on 205–6, "do not know" quote on 211; Green, "Babbage's Analytical Engine." Chambers described the origins of *Vestiges* in an anonymous autobiographical preface added to the tenth revised edition of the book in 1853. It is reprinted in Chambers, *Vestiges*, [204]–[207], "fiats" quote on [204], "organic" quote on [205].

9. Chambers, *Vestiges*, "rustic observer" quote on 165.

10. Ibid., "commencement" quote on 173; Weekes, "Details"; Pocock, "Andrew Crosse"; Secord, *Victorian Sensation*, chap. 5.

11. Chambers, *Vestiges*, "life" quote on 163, "faint representation" quote on 333, "startling idea" quote on 334.

12. Secord, introduction, "vision" quote on xlv, "disarmed" quote on xliv; Secord, *Victorian Sensation*, "best-sellers" quote on 2.

13. Browne, *Charles Darwin Voyaging*, quote on xi; Desmond & Moore, *Darwin*; Charles Darwin to Alexander Humboldt, November 1, 1839 (*Personal*

*Narrative*). Electronic resources for Darwin include *Darwin Correspondence Project*, www.darwinproject.ac.uk/ (source for all letters cited unless otherwise noted) and *The Complete Work of Charles Darwin Online*, edited by John van Wyhe, http://darwin-online.org.uk/ (source for various editions of published work unless otherwise noted).

14. Browne, *Charles Darwin Voyaging*, chap. 1 (chemical lab). Charles Darwin to Caroline S. Darwin, April 8, 1826 (Hope's lectures); C. Darwin, *Beagle Diary*, November 21, 1831 (Harris lecture), July 22, 1832 (St. Elmo's fire), December 4, 1833 (lightning strike); C. Darwin, *Journal of Researches*, January 28, 1832 (cuttlefish) on 6–8.

15. Sarah Elizabeth Wedgwood and Josiah Wedgwood II to Charles Darwin, November 10, 1837 (Crosse); Fara, "Attractive Therapy"; Charles Darwin to William Darwin Fox, December 20, 1844 (mesmerism); Browne, "Retched"; Charles Darwin to Joseph Dalton Hooker, November 5 or 12, 1845 ("quackery" quote), February 8[?], 1846 and July 8 or 15, 1846 (galvanic treatments).

16. Browne, *Charles Darwin Voyaging*; Desmond & Moore, *Darwin*; Joseph Dalton Hooker to Charles Darwin, December 30, 1844 (*Vestiges*); Charles Darwin to Joseph Dalton Hooker, January 7, 1845 (*Vestiges*); Secord, *Victorian Sensation*, chap. 12, "botched" quote on 429; Charles Darwin to William Darwin Fox, April 24[?], 1845 (flattered).

17. C. Darwin, *Foundations*, 128–32; Charles Darwin to Thomas Henry Huxley, December 13, 1856; Lyell, *Principles*, 106; Owen, "Nervous System"; Carpenter, *Principles*, 461–71; A. Wallace, "Fishes"; Timbs, "Electric Eel."

18. C. Darwin, *Natural Selection*, 363–64; Owen, "Nervous System"; Stark, "Existence"; Pauly, *Darwin's Fishes*.

19. C. Darwin, *Origin*, 1st ed., 192–94.

20. Henry Holland to Charles Darwin, December 10, 1859; [Jardine], "Review"; [Wilberforce], "Review."

21. Charles Darwin to Thomas Henry Huxley, November 16, 1860 (M'Donnell); M'Donnell, "On the Organs"; M'Donnell, "On an Organ," quote on 57.

22. Charles Darwin to Charles Lyell, November 24, 1860 (M'Donnell); Charles Darwin to Jeffries Wyman, December 3, 1860 (M'Donnell); Charles Darwin to Thomas Henry Huxley, January 3, 1861 (M'Donnell); Charles Darwin to Jeffries Wyman, February 3, 1861 (M'Donnell); C. Darwin, *Origin*, 4th ed., quotes on 225, 229.

23. Charles Darwin to Asa Gray, July 3, 1860 (design); Charles Darwin to Joseph Dalton Hooker, March 29, 1863 (creation) in *Life and Letters*; Desmond & Moore, *Darwin*, chap. 41; Chadwick, *Secularization*, chap. 7. The rise of the

term "scientist" from the 1860s is clearly seen with the Google Books Ngram viewer at http://ngrams.googlelabs.com/.

24. Campbell, *Reign of Law*; quotes on 103, 108, 109.

25. Lewes, "Mr. Darwin's Hypothesis," 76–78; Charles Darwin to George Henry Lewes, August 7, 1868 in C. Darwin, *More Letters*; A. Wallace, *My Life*, 423–24.

26. Lightman, *Victorian Popularizers*, 47 (Morris); Charles Darwin to George John Romanes, January 24, 1881 in Romanes, *Life and Letters*, 106–7 (parable).

27. Glotzhaber et al., *Student Laboratory Manual*, quote on 55; Wieland, "Ghostly Coincidence"; Thomas, "Is There Evolution"; Edis, "Grand Themes"; for Yahya, see, e.g., www.evidencesofcreation.com/nature04.htm.

28. Knight, *Humphry Davy*; H. Davy, "Account," quotes on 16, 17; J. Davy, "Account."

29. J. Davy, "Account," quotes on 260, 276; J. Davy, "Observations"; Mottelay, *Bibliographical History*, 344, 477–78.

30. Müller, "Phenomena"; Piccolino & Bresadola, "Drawing a Spark"; Otis, "Metaphoric Circuit"; Clarke & Jacyna, *Nineteenth-Century Origins*, chap. 5; Linari, "Inquiry." Colladon's galvanometer subsequently ended up in the South Kensington Museum; see the *Catalogue* of 1876.

31. Faraday, "Notice," Humboldt quoted on 2–3; B. Jones, *Life and Letters*, Faraday quoted on 84; Timbs, "Electric Eel"; Bradley, "Electric Eel"; Schwarz, "Faraday and Babbage"; Altick, *Shows*, 379; Lightman, *Victorian Popularizers*. For captive electric eels, see Fahlberg, "Description" (Stockholm); "Natural History. Electric Eel" (electric eel in Paris); *History of Paris*, 2:441–42 (ditto).

32. Timbs, "Electric Eel," "waistcoats" quote on 35; Allen, *Philosophy*, "bold life-guardsman" quote on 761; Morus, "Electric Ariel"; Morus, "Currents"; Schwarz, "Faraday and Babbage"; Hirshfeld, *Electric Life*, Faraday "recluse" quote on 150; Charles Babbage to Michael Faraday, August 14, 1839; Michael Faraday to Charles Babbage, August 15, 1839; Michael Faraday to Angela Georgina Burdett Coutts, August 15, 1839; Michael Faraday to Angela Georgina Burdett Coutts, May 23, 1840, all in Faraday, *Correspondence*, 599–600, 672. The life-guardsman episode is also mentioned in Smee's 1849 *Elements of Electro-biology*, 89: "the stoutest lifeguardsman has been known to faint when he has been acted upon by the shock."

33. Hirshfeld, *Electric Life*, quote on 101.

34. Ibid., quote on 147; Piccolino, "Taming"; Gee, "Early Development"; Mills, "Early Voltaic Batteries" (mercury poisoning).

35. Faraday, "Notice," quote on 1; Faraday, "Identity"; Herschel, *Discourse*, 341–43.

36. Faraday, "Notice," quotes on 3, 4; Daniell, "Animal Electricity"; Owen, "Nervous System," quote on 217; Morus, "Currents," "elite" quote on 53; Timbs, "Electric Eel"; Reynolds, "Todd."

37. Faraday, "Notice," quotes on 4, 6; Schaffer, "Self Evidence," quote on 330.

38. Faraday, "Notice"; Daniell, "Animal Electricity," quote on 502.

39. Faraday, "Notice," quotes on 1, 11; Daniell, "Animal Electricity," quote on 502.

### *Chapter Six. Electric Currents*

1. Clarke & Jacyna, *Nineteenth-Century Origins*, chap. 5, quote on 159; Hoff & Geddes, "Rheotome."

2. Clarke & Jacyna, *Nineteenth-Century Origins*, chap. 5; Hoff & Geddes, "Rheotome"; Piccolino, "Animal Electricity"; Brazier, *19th Century*, chap. 3.

3. Hoff & Geddes, "Rheotome."

4. Clarke & Jacyna, *Nineteenth-Century Origins*, chap. 5; Brazier, *19th Century*, chap. 3; Moruzzi, "Electrophysiological Work," Matteucci quoted on 137 (Moruzzi's translation); Matteucci, "Electro-Physiological Researches."

5. Hoff & Geddes, "Rheotome," Matteucci's "facts" quote on 219 (their translation); Clarke & Jacyna, *Nineteenth-Century Origins*, chap. 5, "revealed" quote on 198, Matteucci's "sensitive apparatus" quote on 199; Moruzzi, "Electrophysiological Work"; Antolini, "Seeing the Heartbeat"; Geddes, "Capillary Electrometer."

6. Matteucci, "Electro-Physiological Researches, Second Memoir," "fowls" quote on 300; Matteucci, "Electro-Physiological Researches, Fourth Memoir"; Brazier, *19th Century*, chap. 3; Clarke & Jacyna, *Nineteenth-Century Origins*, chap. 5.

7. Matteucci, "Electro-Physiological Researches, Fourth Memoir," "certain" quote on 485; Clarke & Jacyna, *Nineteenth-Century Origins*, chap. 5; Brazier, *19th Century*, chaps. 5 & 6; Otis, *Müller's Lab*, chap. 1; Cranefield, "Organic Physics."

8. Clarke & Jacyna, *Nineteenth-Century Origins*, chap. 5; Brazier, *19th Century*, chaps. 5 & 6; Otis, *Müller's Lab*, chap. 1, Müller's "quality" quote on 9, "intrigued" quote on 23; Cranefield, "Philosophical."

9. Cranefield, "Organic Physics"; Clarke & Jacyna, *Nineteenth-Century Origins*, chap. 5, quote on 159; Hoff & Geddes, "Rheotome."

10. Otis, *Müller's Lab*, chaps. 1 & 3, Du Bois-Reymond's "vermin" quote on 76; Brazier, *19th Century*, chap. 6; Otis, "Metaphoric Circuit"; Hoff & Geddes, "Rheotome"; Lenoir, "Models."

11. Dierig, "Urbanization," Du Bois-Reymond's "martyr" quote on 5; Brazier, *19th Century*, chaps. 5 & 6, "wherever" quote on 74; Otis, *Müller's Lab*, chap. 3, "walking" quote on 90; Finkelstein, "Du Bois-Reymond," "writhing" quote on 27 (his translation).

12. Otis, *Müller's Lab*, chaps. 1 & 3; Otis, "Metaphoric Circuit"; Finkelstein, "Du Bois-Reymond"; Hoff & Geddes, "Rheotome."

13. Otis, *Müller's Lab*, chap. 3, quote on 91; Hoff & Geddes, "Rheotome"; Geddes, "Did Wheatstone."

14. Otis, *Müller's Lab*, "tension" quote on 5,

15. Ibid.; Otis, "Metaphoric Circuit"; Lenoir, "Helmholtz," quote on 185; Pouillet, "Note"; Hoff & Geddes, "Graphic Recording."

16. Otis, *Müller's Lab*; Otis, "Metaphoric Circuit," Du Bois-Reymond quoted on 105 (her translation), 114; Lenoir, "Helmholtz"; Verkhratsky et al., "Galvani."

17. T. Kuhn, "Energy Conservation," quotes on 321.

18. Ibid.

19. Morus, "Manufacturing Nature"; Morus, "Nervous System"; Geddes, "Did Wheatstone."

20. Morus, "'Nervous System'"; Otis, "Metaphoric Circuit"; Otis, *Networking*; Geddes & Geddes, "Georg Simon Ohm"; Geddes, "Did Wheatstone."

21. Morus, *Physics*, chap. 4; Morus, *Frankenstein's Children*, chap. 7; Morus, "Nervous System," Timbs quote on 471; Kirkland, "High-Tech Brains." A search of Google Books (June 3, 2011) shows that the quote from the Book of Job appeared in at least a hundred nineteenth-century discussions of the telegraph.

22. Grove, "Address," quote on 486.

23. Lenoir, "Helmholtz"; Grove, *Correlation*; Cantor, "William Robert Grove"; Green, "Babbage"; Carpenter, *Nature and Man*, quote from letter on 50, "automatism" essay on 261–83, quotes on 262, 268; Morus, "Nervous System," Carpenter quoted on 471–73; Hall, "Contribution."

24. Morus, *Physics*, chap. 4; Morus, *Frankenstein's Children*, chap. 7; Morus, "Nervous System"; Griffiths, *Introduction*, Hertz quote on xiii; Feynman, *Lectures*, 1:2-2; Mahon, *Man*; Hunt, *Maxwellians*.

25. Fraden, *Handbook*; Hope, "100 Years." The utility of the galvanometer for scientific measurements of various kinds can still be seen in books written for amateur/citizen scientists, like Forrest M. Mims III's *Science and Communication Circuits* and *Electronic Sensor Circuits*.

26. Otis, *Müller's Lab*; Pouillet, "Note," quote on 568; Hoff & Geddes, "Rheotome"; Finkelstein, "Du Bois-Reymond."

27. Du Bois-Reymond, "Observations"; Goodsir, "Anatomical Details"; Murray, "On Electrical Fishes"; Murray, "Remarks," "abnegation" quote on 159.

28. Du Bois-Reymond, "Observations," "touched" quote on 387, "coffin" quote on 382.

29. Ibid., "violent tumult" quote on 376–77. For techno-organic hybrids of various sorts, see Dierig, "Engines"; Channell, *Vital Machine.*

30. Dierig, "Urbanization," Du Bois-Reymond's quote on 7.

31. Ibid.

32. Otis, *Müller's Lab*, chap. 3, quote on 88. Moller, *Electric Fish*, 35, reports that Sachs "humbly noted" that Humboldt's story about fishing for horses with electric eels "seemed to be somewhat exaggerated."

33. Caton, "Electric Currents"; Brazier, "History"; Brazier, "Pioneers"; Geddes, "Did Wheatstone," "beam" quote on 89; Geddes, "What."

34. Otis, "Howled," trial charge quoted on 27; Pedlar, "Experimentation"; Public General Act, 1879, 39 & 40 Vict., c. 77; Breathnach, "Eduard Hitzig"; Grundfest, "Different Careers"; Brazier, *19th Century*, chap. 10; Clarke & Jacyna, *Nineteenth-Century Origins*, chap. 6; Morgan, "First Reported Case."

35. Harris & Almerigi, "Probing," Bartholow quotes on 101–2; Boling et al., "Historical Contributions," Bartholow's "criminal" quote on 1299; Morgan, "First Reported Case"; Lederer, *Subjected*, chap. 1; Bartholow, *Medical Electricity.*

36. Grundfest, "Different Careers."

37. Ibid.; Dierig, "Urbanization"; Dierig, "Engines," "piece" quote on 125; Channell, *Vital Machine.*

38. Du Bois-Reymond, "Observations," "Adriatic" quote on 418; Grundfest, "Different Careers"; Dierig, "Urbanization."

39. Du Bois-Reymond, "Observations"; Grundfest, "Different Careers"; Dierig, "Urbanization"; Brazier, *19th Century.*

40. Dierig, "Engines," quote on 130; Marey, *Animal Mechanism*; Hoff et al., "Anemograph"; Burnett, "Origins"; Borrell, "Instrumentation." Hoff and colleagues show that a clockwork kymograph-like device was used for monitoring wind direction and velocity as early as 1734.

41. Marey, *Animal Mechanism*; Douard, "E.-J. Marey"; Borrell, "Instrumentation," quote on 53; Brain, "Pulse"; Brazier, *19th Century*, chap. 10.

42. Marey, *Animal Mechanism*; Geddes & Hoff, "Capillary Electrometer"; Burnett, "Origins"; Douard, "E.-J. Marey"; Borrell, "Instrumentation."

43. Moller, *Electric Fish*; Borrell, "Instrumentation"; Dombois, "Muscle Telephone"; Volmar, "Listening," Bernstein quoted on 8; Kirkland, "High-Tech Brains." For Bell, see the Library of Congress American Treasures exhibit at www.loc.gov/exhibits/treasures/trr002.html, accessed June 15, 2011. Dombois uses "sonification" to refer to the general process of portraying data as sounds

(parallel to visualization) and "audification" to refer to the specific process of translating a data waveform directly into sound.

44. Volmar, "Listening," Wedensky quote on 10; Moller, *Electric Fish*; Chambers Journal, "Curiosities"; Burdon Sanderson & Gotch, "Electrical Organ," quote on 141; Holder, "Electricians"; Dombois, "Muscle Telephone."

45. Douard, "E.-J. Marey"; Daston & Galison, "Image," Marey quoted on 81; Brain, "Pulse," quote on 402.

### *Chapter Seven. Discovering Electric Worlds*

1. Hughes, *Sensory Exotica*; Uexküll, *Foray*.

2. Piccolino, "Luigi Galvani."

3. Hoff & Geddes, "Rheotome"; Schuetze, "Discovery"; Seyfarth, "Julius Bernstein"; Frank, "Instruments"; Piccolino, "Luigi Galvani."

4. Moller, *Electric Fish*; Frank, "Instruments," quotes on 211, 213.

5. Seyfarth, "Bernstein"; Verkhratsky et al., "From Galvani"; Leuchtag, "Animal Electricity"; Piccolino, "Luigi Galvani."

6. Bradley & Tansey, "Coming"; Frank, "Instruments"; Finger, "Edgar D. Adrian"; Piccolino, "Luigi Galvani"; Piccolino and Bresadola, "Drawing a Spark." The "all-or-none" principle was renamed "all-or-nothing" in 1922, Brazier, "Historical Development," 24.

7. Burnett, "Origins," quote on 70; Bradley & Tansey, "Coming"; Frank, "Instruments," "weighed" quote on 212; Anderson, "Story."

8. Bradley & Tansey, "Coming"; Frank, "Instruments"; Finger, "Edgar D. Adrian"; Burnett, "Origins"; Moller, *Electric Fish*, 24–29.

9. Finger, "Edgar D. Adrian"; Frank, "Instruments."

10. Frank, "Instruments"; Finger, "Edgar D. Adrian," Wiener, *Cybernetics*, 182.

11. Bradley & Tansey, "Coming," Adrian "valve" quote on 221; Frank, "Instruments," Adrian "pleased" quote on 229; Finger, "Edgar D. Adrian."

12. Frank, "Instruments"; Bradley & Tansey, "Coming"; Geddes & Hoff, "Capillary Electrometer."

13. Frank, "Instruments," quote on 209; Barbara, "Fessard's School"; Barbara, "Franco-British Relations"; Whittaker, "Arcachon"; Whittaker, "Historical Significance."

14. Jacobson, *Foundations*, 35; J. Robinson, *Mechanisms*, Cajal quote on 21; Brazier, "Historical Development," Elliott quote on 25; Bennett, "One Hundred Years"; Feldberg & Fessard, "Cholinergic Nature."

15. Keesey, "How"; Moller, *Electric Fish*, 33–34; Whittaker, "Arcachon"; Whittaker, "Historical Significance"; Barbara, "Fessard's School"; Barbara, "Franco-British Relations"; Whittaker, "Biochemistry."

16. Moller, *Electric Fish.*

17. Ibid.; "General Notes."

18. Moller, *Electric Fish*; Alexander, "Lissmann"; Uexküll, *Foray*, quotes on 42, 51; Burkhardt, *Patterns*, Ch. 3.

19. Moller, *Electric Fish*; Alexander, "Lissmann."

20. Moller, *Electric Fish*; Lissmann, "Continuous Electrical Signals," 201; Lissmann, "On the Function," 157; Alexander, "Lissmann," 242.

21. Moller, *Electric Fish*; Peters, "Academic Freedom"; Cox, "Electric Fish."

22. Au, *Sonar*, 2–3; Hughes, *Sensory Exotica*; Frank, "Instruments," "neurons" quote on 210.

23. Horowitz & Hill, *Art of Electronics*, quote on 4; Mindell, *Between Human and Machine.*

24. Haring, *Ham Radio's Technical Culture*; Hintz, "Portable Power," "forced" quote by Philip Rogers Mallory on 34; Riordan & Hoddeson, *Crystal Fire*; Seitz & Einspruch, *Electronic Genie.*

25. Moller, *Electric Fish*, Lissmann quote from the foreword on xviii; Lissmann, "Function," quotes on 158, 159; T. Roberts, "Review."

26. Lissmann, "Function," quotes on 160, 165.

27. Lissmann, "James Gray," 63–64; Alexander, "New Sense"; Olshanckiy, "Body-Size Electric Eye." Lissmann & Machin, "Mechanism," quote on 457.

28. Ibid.; Alexander, "New Sense," office mate quoted on 201.

29. Lissmann & Machin, "Mechanism."

30. Machin & Lissmann, "Mode."

31. "Electric Eel," *Time*, April 26, 1937; Coates, "Kick."

32. Lissmann, "On the Function," 186.

33. Moller, *Electric Fish*; Bullock, "Electroreception"; Bodznick & Montgomery, "Physiology," quote on 132; Zupanc & Bullock, "From Electrogenesis"; Rose, "Insights"; Pettigrew, "Electroreception."

34. Kalmijn, "Electro-perception"; Kalmijn, "Electric Sense"; Zupanc & Bullock, "From Electrogenesis"; Bullock, "Electroreception," quote on 127.

35. Zupanc & Bullock, "From Electrogenesis"; Hopkins, "Design Features"; Hopkins, "Signal Evolution." Bullock, "Electroreception," quote on 126; Moller, *Electric Fish*, 96–100.

36. Cressey, "Please"; Albert et al., "Case," quote on 347; Keesey, "How"; Lindstrom "Cause"; Folgering & Poolman, "Channel Electrophysiology"; Xu & LaVan, "Designing."

37. Bullock et al., *Electroreception*; Bullock & Hopkins, "Explaining Electroreception"; Moller, *Electric Fishes.*

38. Bullock, "Electroreception"; Heiligenberg & Bastian, "Electric Sense";

Ingle & Crews, "Vertebrate Neuroethology," quote on 458; Carr, "Neuroethology"; Hopkins, "Neuroethology"; Hopkins, "Convergent Design."

39. Kirkland, "High-Tech Brains"; Walter, *Living Brain*; Harmon & Lewis, "Neural Modeling"; Platt, "Amplification"; Taylor, "Computers"; Pickering, *Cybernetic Brain*.

40. Lorenz, "Comparative Method"; Lorenz, *Foundations*, "complication" quotes on 90, 20; Burkhardt, *Patterns*. Inductors are the opposites of capacitors: "the rate of current change in an inductor depends on the voltage applied across it, whereas the rate of voltage change in an inductor depends on the current through it." Horowitz & Hill, *Art of Electronics*, quote on 28.

41. Heiligenberg, *Neural Nets*; Heiligenberg & Bastian, "Electric Sense," quote on 562; Bullock, "Electroreception"; Rose, "Insights."

42. Heiligenberg, *Neural Nets*; Rose, "Insights."

43. Heiligenberg, *Neural Nets*; Bullock, "Electroreception"; Hopkins, "Convergent Designs."

44. Alves-Gomes, "Evolution"; Carlson et al., "Brain Evolution"; Rose, "Insights"; Hopkins, "Convergent Designs"; Hopkins, "Design Features"; Hopkins, "Signal Evolution"; Moller, *Electric Fish*.

45. Laland et al., "Learning in Fishes," "obsolete" quote on 199; Laland & Hoppitt, "Do Animals Have Culture?"; Bshary et al., "Fish Cognition"; Brown & Laland, "Suboski"; Kuba et al., "New Method"; Shubin, *Your Inner Fish*, quote in ix.

### *Conclusion. Nothing but a Movement of Electrons*

1. Chaisson, *Epic*, quote on 54; Christian, *Maps of Time*.

2. Christian, *Maps of Time*; Chaisson, *Epic*; Darwin to Joseph Dalton Hooker, in C. Darwin, *Life and Letters*, 18n.

3. Christian, *Maps of Time*; Chaisson, *Epic*; Luisi, *Emergence*; Chaisson, *Cosmic Evolution*; Goodsell, *Machinery*.

4. Szent-Györgyi in McElroy & Glass, *Symposium*, "life" quotes on 7; Goodsell, *Machinery*, "control" quote on 42.

5. Cooper & Schliwa, "Motility"; Chaisson, *Epic*.

6. Goodsell, *Machinery*; Vogel, *Cats' Paws*, quotes on 162, 163.

7. Stevens, "Light-at-night"; Hayes, *Infrastructure*; Russell, *Evolutionary History*.

8. Riordan & Hoddeson, *Crystal Fire*; Moore, "Cramming"; Cressler, *Silicon Earth*. The annual amount of precipitation in the coterminous United States is approximately 1,430 cubic miles of water, or ~$5.96 \times 10^{18}$ cubic centimeters; USGS Washington Water Science Center, "Rain." If we assume a raindrop has a

radius of 0.2 cm (and volume of 0.034 cubic centimeters), there are about 29.4 raindrops in a cubic centimeter of water, and thus about $17.5 \times 10^{19}$ raindrops in total.

9. Christian, *Maps of Time*, quote on 446; McNeill & McNeill, *Human Web*; Lyman & Varian, "How Much Information"; Bohn & Short, "How Much Information."

10. Otis, *Networking*; DeLanda, *Philosophy*; Floreano & Mattiussi, *Bio-Inspired Artificial Intelligence*; Shasha & Lazere, *Natural Computing*; Bray, *Wetware*; Sarpeshkar, *Ultra Low Power Bioelectronics*.

11. Horowitz & Hill, *Art of Electronics*, quotes on 3; "Number of Internet Users to Surpass 2 Billion by End of Year, UN Agency Reports," UN News Centre (October 19, 2010); "Mobile Telephones More Common than Toilets in India, UN Report Finds," UN News Centre (April 14, 2010).

# Bibliography


Abd al-Latif. *Relation de L'Egypte, par Abd-Allatif, médecin arabe de Bagdad.* Translated by Silvestre de Sacy. Paris: Treuttel & Würtz, 1810.

"An Act to Amend the Law Relating to Cruelty to Animals." Public General Act, 39 & 40 Victoria, C. 77. 1876. http://www.animalrightshistory.org/animal-rights-law/victorian-legislation/1876-uk-act-vivisection.htm.

Adams, W. Grylls. "On Grove's, Plante's, and Faure's Secondary Batteries." *Chemical News and Journal of Physical Science* 45, no. 1154 (1882): 1–5.

Aegineta, Paulus. *The Medical Works of Paulus Aegineta, the Greek Physician.* Vol. 1. Translated by Francis Adams. London: J. Welsh, 1834.

———. *The Seven Books of Paulus Aegineta.* Edited by Francis Adams. Translated by Francis Adams. London: Sydenham Society, 1847.

Agamben, Giorgio. *The Open: Man and Animal.* Translated by Kevin Attell. Stanford, CA: Stanford University Press, 2003.

Agazzi, Elena. "The Impact of Alessandro Volta on German Culture." In *Nuova Voltiana: Studies on Volta and His Times*, vol. 4, edited by Fabio Bevilacqua and Lucio Fregonese. Milan: Ulrico Hoepli, 2002.

Albert, J. S., H. H. Zakon, P. K. Stoddard, A. Unguez, S. K. S. Holmberg-Albert, and M. R. Sussman. "The Case for Sequencing the Genome of the Electric Eel *Electrophorus Electricus*." *Journal of Fish Biology* 72, no. 2 (2008): 331–54.

Aldini, Giovanni. *Essai théorique et expérimental sur le galvanisme, avec une série d'expériences faites en présence des commissaires de L'Institut National de France, et en divers amphithéâtres anatomiques de Londres par Jean Aldini.* Paris: De l'Imprimerie de Fournier Fils, 1804.

Alexander, R. McNeill. "Hans Werner Lissmann, 30 April 1909–21 April 1995." *Biographical Memoirs of Fellows of the Royal Society* 42 (1996): 234–45.

———. "A New Sense for Muddy Water." *Journal of Experimental Biology* 209 (2006): 200–201.

———. "Structure and Function in the Catfish." *Journal of Zoology* 148, no. 1 (1965): 88–152.

Allen, Zachary. *Philosophy of the Mechanics of Nature, and the Source and Modes of Action of Natural Motive-Power*. New York: D. Appleton, 1852.

Althaus, Julius. *A Treatise on Medical Electricity, Theoretical and Practical, and Its Use in the Treatment of Paralysis, Neuralgia, and Other Diseases*. 2nd ed. Philadelphia: Lindsay & Blakiston, 1870.

Altick, Richard D. *The Shows of London*. Cambridge, MA: Harvard University Press, 1978.

Alves-Gomes, J. A. "The Evolution of Electroreception and Bioelectrogenesis in Teleost Fish: A Phylogenetic Perspective." *Journal of Fish Biology* 58 (2001): 1489–1511.

Ambrose, Stanley H. "Paleolithic Technology and Human Evolution." *Science* 291, no. 5509 (2001): 1748–53.

Andel, Tjeerd H. van. "Late Pleistocene Sea Levels and the Human Exploitation of the Shore and Shelf of Southern South Africa." *Journal of Field Archaeology* 16, no. 2 (1989): 133–55.

Anderson, Antony. "The Story of Cambridge Instruments [Review of Cattermole & Wolfe, Horace Darwin's Shop]." *New Scientist* 114, no. 1564 (1987): 57–58.

Andrews, P. "Evolution and Environment in the Hominoidea." *Nature* 360, no. 6405 (1992): 641–46.

Antolini, R. "Seeing the Heartbeat: From Matteucci's Rheoscopic Frog to Image-Guided Arrhythmia Ablation." *Il nuovo saggiatore* 26, no. 1–2 (2010).

Apicus. *Cookery and Dining in Imperial Rome*. Edited and translated by Joseph Dommers Vehling. New York: Dover, 1977.

Appel, Toby A. "Geoffroy and the Emergence of Philosophical Anatomy." In *The Cuvier-Geoffroy Debate: French Biology in the Decades before Darwin*. New York: Oxford University Press, 1987.

Aristotle. *History of Animals*. Translated by Richard Cresswell. London: George Bell & Sons, 1883.

———. *On the Parts of Animals*. Translated by W. Ogle. London: Kegan Paul, Trench, 1882.

Arthur, W. Brian. *The Nature of Technology: What It Is and How It Evolves*. New York: Free Press, 2009.

Asúa, Miguel de. "The Experiments of Ramoacuten M. Termeyer SJ on the Electric Eel in the River Plate Region (c. 1760) and Other Early Accounts of *Electrophorus Electricus*." *Journal of the History of the Neurosciences* 17, no. 2 (2008): 160–75.

Au, Whitlow W. L. *The Sonar of Dolphins*. New York: Springer, 1993.

Bailey, Geoff, José S. Carrión, Darren A. Fa, Clive Finlayson, Geraldine Finlayson, and Joaquín Rodríguez-Vidal. "The Coastal Shelf of the Mediterranean and Beyond: Corridor and Refugium for Human Populations in the Pleistocene." *Quaternary Science Reviews* 27, no. 23–24 (2008): 2095–99.

Baird, William. *Cyclopaedia of the Natural Sciences*. London: Richard Griffin, 1858.

Bancroft, Edward. *An Essay on the Natural History of Guiana in South America*. London: Printed for T. Becket and P. A. De Hondt, 1769.

Barbara, J. G. "The Fessard's School of Physiology after War in France: Globalization and Diversity in Neurophysiological Research on Torpedo Fish (1938–1955)." Presented at the 22nd International Conference for the History of Science, IUHPS/DHS, Beijing, 2005.

———. "Franco-British Relations in Neurophysiology in Edgar Adrian's Era." Presented at the Second International Colloquium of the European Society for the History of Science and la Société Française d'Histoire des Sciences et des Techniques: "Echanges Franco-Britanniques entre savants depuis le XVIIe siècle," Maison Française d'Oxford, 2006.

Bartholow, Roberts. *Medical Electricity: A Practical Treatise on the Applications of Electricity to Medicine and Surgery*. Philadelphia: Henry C. Lea's Son, 1882.

Beard, George Miller, and Alphonso David Rockwell. *A Practical Treatise on the Medical and Surgical Uses of Electricity*. 4th ed. New York: William Wood, 1883.

Behlmer, George K. "Grave Doubts: Victorian Medicine, Moral Panic, and Signs of Death." *Journal of British Studies* 42, no. 2 (2003): 206–35.

Behn, Aphra. *Oroonoko, and Other Writings*. Oxford: Oxford University Press, 1998.

Belbenoit, Pierre, Peter Moller, Jacques Serrier, and Stephen Push. "Ethological Observations on the Electric Organ Discharge Behaviour of the Electric Catfish, *Malapterurus Electricus* (Pisces)." *Behavioral Ecology and Sociobiology* 4, no. 4 (1979): 321–30.

Ben-Amos, Paula. "Men and Animals in Benin Art." *Man* 2, no. 2 (1976): 243–52.

Bennett, M. R. "One Hundred Years of Adrenaline: The Discovery of Autoreceptors." *Clinical Autonomic Research* 9, no. 3 (1999): 145–59.

Benyus, Janine M. *Biomimicry: Innovation Inspired by Nature*. New York: Harper Perennial, 1997.

Bernardi, Walter. "The Controversy on Animal Electricity in Eighteenth-Century Italy: Galvani, Volta, and Others." In *Nuova Voltiana: Studies on Volta and His Times*, vol. 1, edited by Fabio Bevilacqua and Lucio Fregonese. Milan: Ulrico Hoepli, 2000.

Berry, A. J. "Some Aspects of Early Electrochemistry." *Bulletin of the British Society for the History of Science* 1, no. 8 (1952): 205–10.

Bertucci, Paola. "Marco Piccolino; Marco Bresadola. *Rane, Torpedini e Scintille: Galvani, Volta e L'elettricità Animale.*" *Isis* 96, no. 2 (2005): 284–85.

———. "The Shocking Bag: Medical Electricity in Mid-18th-Century London." In *Nuova Voltiana: Studies on Volta and His Times*, vol. 5, edited by Fabio Bevilacqua and Lucio Fregonese. Milan: Ulrico Hoepli, 2003.

———. "Sparks in the Dark: The Attraction of Electricity in the Eighteenth Century." *Endeavour* 31, no. 3 (2007): 88–93.

———. "Therapeutic Attractions: Early Applications of Electricity to the Art of Healing." In Whitaker, Smith, and Finger, *Brain, Mind, and Medicine.*

Bichat, Xavier. *Physiological Researches upon Life and Death.* 1st American ed. Philadelphia, PA: Smith & Maxwell, 1809.

Blier, Suzanne Preston. *The Royal Arts of Africa: The Majesty of Form.* Upper Saddle River, NJ: Prentice Hall, 2003.

Bodznick, David, and John C. Montgomery. "The Physiology of Low-Frequency Electrosensory Systems." In *Electroreception.* Edited by Theodore H. Bullock, Carl D. Hopkins, Arthur N. Popper, and Richard R. Fay. New York: Springer, 2005.

Boesch, Christophe. "Is Culture a Golden Barrier Between Human and Chimpanzee?" *Evolutionary Anthropology* (2003): 82–91.

Boesch, Christophe, and Hedwige Boesch. "Mental Maps in Chimpanzees: An Analysis of Hammer Transport for Nut Cracking." *Primates* 25, no. 2 (1984): 160–70.

Bohn, Roger E., and James E. Short. How Much Information? Report on American Consumers, San Diego: University of California. Global Information Industry Center, 2009. http://hmi.ucsd.edu/pdf/HMI_2009_ConsumerReport_Dec9_2009.pdf.

Boling, Warren, Andre Olivier, and Gavin Fabinyi. "Historical Contributions to the Modern Understanding of Function in the Central Area." *Neurosurgery* 50, no. 6 (2002): 1296–1310.

Borell, Merriley. "Instrumentation and the Rise of Physiology." *Science and Technology Studies* 5, no. 2 (1987): 53–62.

Borenstein, Elhanan, Marcus W. Feldman, and Kenichi Aoki. "Evolution of Learning in Fluctuating Environments: When Selection Favors Both Social and Exploratory Individual Learning." *Evolution* 62, no. 3 (2008): 596–602.

Bradley, J. K., and E. M. Tansey. "The Coming of the Electronic Age to the Cambridge Physiological Laboratory: E. D. Adrian's Valve Amplifier in 1921." *Notes and Records of the Royal Society of London* 50, no. 2 (1996): 217–28.

Bradley, Thomas. "Electric Eel at the Adelaide Gallery." *Charlesworth's Magazine of Natural History* 3 (1839): 564–65.

Brain, Robert Michael. "Ørsted's Concept of Force and Theory of Music." In Brain, Cohen, and Knudsen, *Hans Christian Ørsted.*

———. "The Pulse of Modernism: Experimental Physiology and Aesthetic Avant-Gardes Circa 1900." *Studies in History and Philosophy of Science* 39, no. 3 (2008): 393–417.

Brain, Robert M., Robert S. Cohen, and Ole Knudsen. *Hans Christian Ørsted and the Romantic Legacy in Science.* Boston Studies in the Philosophy of Science 241. Dordrecht, NL: Springer, 2007.

Braun, David R., John W. K. Harris, Naomi E. Levin, Jack T. McCoy, Andy I. R. Herries, Marion K. Bamford, Laura C. Bishop, Brian G. Richmond, and Mzalendo Kibunjia. "Early Hominin Diet Included Diverse Terrestrial and Aquatic Animals 1.95 Ma in East Turkana, Kenya." *Proceedings of the National Academy of Sciences of the United States of America* 107, no. 22 (2010): 10002–7.

Bray, Dennis. *Wetware: A Computer in Every Living Cell.* New Haven, CT: Yale University Press, 2011.

Bray, Richard N., and Mark A. Hixon. "Night-Shocker: Predatory Behavior of the Pacific Electric Ray (*Torpedo Californica*)." *Science* 200, no. 4339 (1978): 333–34.

Brazier, Mary A. B. "The Historical Development of Neurophysiology." In *Handbook of Physiology, Section 1: Neurophysiology*, vol. 1., edited by I. J. Field. Washington, DC: American Physiological Society, 1959.

———. *A History of Neurophysiology in the 17th and 18th Centuries: From Concept to Experiment.* New York: Raven Press, 1984.

———. *A History of Neurophysiology in the 19th Century.* New York: Raven Press, 1988.

———. "The History of the Electrical Activity of the Brain as a Method for Localizing Sensory Function." *Medical History* 7, no. 3 (1963): 199–211.

———. "Pioneers in the Discovery of Evoked Potentials." *Electroencephalography and Clinical Neurophysiology* 59, no. 1 (1984): 2–8.

Breathnach, Caoimhghin S. "Eduard Hitzig, Neurophysiologist and Psychiatrist." *History of Psychiatry* 3, no. 11 (1992): 329–38.

Bresadola, Marco. "At Play with Nature: Luigi Galvani's Experimental Approach to Muscular Physiology." In *Reworking the Bench: Research Notebooks in the History of Science*, edited by Frederic L Holmes, Jürgen Renn, and Hans-Jörg Rheinberger. Dordrecht: Kluwer, 2003.

———. "Medicine and Science in the Life of Luigi Galvani (1737–1798)." *Brain Research Bulletin* 46, no. 5 (1998): 367–80.

Brier, B., and M. V. L. Bennett. "Autopsies on Fish Mummies: Possible Identification of the Classical Phagrus." *Journal of Egyptian Archaeology* 65 (1979): 128–33.

Briggs, John C. "The Biogeography of Otophysan Fishes (Ostariophysi: Otophysi): A New Appraisal." *Journal of Biogeography* 32, no. 2 (2005): 287–94.

Bronk Ramsey, Christopher, Michael W. Dee, Joanne M. Rowland, Thomas F. G. Higham, Stephen A. Harris, Fiona Brock, Anita Quiles, Eva M. Wild, Ezra S. Marcus, and Andrew J. Shortland. "Radiocarbon-Based Chronology for Dynastic Egypt." *Science* 328, no. 5985 (2010): 1554–57.

Brown, C., and K. N. Laland. "Suboski and Templeton Revisited: Social Learning and Life Skills Training for Hatchery Reared Fish." *Journal of Fish Biology* 59 (2001): 471–93.

Browne, Janet. *Charles Darwin Voyaging: A Biography*. Princeton, NJ: Princeton University Press, 1995.

———. "I Could Have Retched All Night: Charles Darwin and His Body." In *Science Incarnate: Historical Embodiments of Natural Knowledge*, edited by Christopher Lawrence and Steven Shapin. Chicago: University of Chicago Press, 1998.

Brown-Sequard, C. E. "The Croonian Lecture: On the Relations between Muscular Irritability, Cadaveric Rigidity, and Putrefaction." *Proceedings of the Royal Society of London* 11 (1860): 204–14.

Bryant, William. "Account of an Electrical Eel, or the Torpedo of Surinam." *Transactions of the American Philosophical Society* 2 (1786): 166–69.

Bshary, Redouan, Wolfgang Wickler, and Hans Fricke. "Fish Cognition: A Primate Eye's View." *Animal Cognition* 5 (2002): 1–13.

Builth, Heather. "Gunditjmara Environmental Management: The Development of a Fisher-Gatherer-Hunter Society in Temperate Australia." In *Beyond Affluent Foragers: Rethinking Hunter-Gatherer Complexity*, edited by Colin Grier, Jangsuk Kim, and Junzo Uchiyama. Oxford: Oxbow Books, 2006.

Bullock, T. H. "Electroreception." *Annual Review of Neuroscience* 5 (1982): 121–70.

Bullock, Theodore H., and Carl D. Hopkins. "Explaining Electroreception." In Bullock et al., *Electroreception*.

Bullock, Theodore H., Carl D. Hopkins, Arthur N. Popper, and Richard R. Fay, eds. *Electroreception*. New York: Springer, 2005.

Burdon Sanderson, J., and Francis Gotch. "On the Electrical Organ of the Skate." *Journal of Physiology* 9, no. 2–3 (1888): 137–66.

———. "On the Electrical Organ of the Skate, Part 2." *Journal of Physiology* 10 (1889): 259–328.

Burkhardt, Richard W, Jr. *Patterns of Behavior: Konrad Lorenz, Niko Tinbergen, and the Founding of Ethology*. Chicago: University of Chicago Press, 2005.

Burnett, John. "The Origins of the Electrocardiograph as Clinical Instrument." *Medical History* supplement no. 5 (1985): 53–76.

Byrne, R. W. "The Manual Skills and Cognition That Lie behind Hominid Tools Use." In *The Evolution of Thought*, edited by A. E. Russon and D. R. Begun. Cambridge: Cambridge University Press, 2004.

California Department of Fish and Game. "Skates and Rays." In *California's Marine Living Resources: A Status Report*. December 2001. www.dfg.ca.gov/marine/status/skates_and_rays.pdf.

———. "Skates and Rays." In *Status of the Fisheries Report: An Update through 2008*. August 2010. www.dfg.ca.gov/marine/status/report2008/skates.pdf.

Campbell, George. *The Reign of Law*. London: Alexander Strahan, 1867.

Canton, John. "A Letter to the Right Honourable the Earl of Macclesfield, President of the Royal Society, Concerning Some New Electrical Experiments." *Philosophical Transactions (1683–1775)* 48 (1753): 780–85.

Cantor, G. N. "William Robert Grove, the Correlation of Forces, and the Conservation of Energy." *Centaurus* 19, no. 4 (1975): 273–90.

Caputi, Angel A. "Contributions of Electric Fish to Understanding Sensory Processing by Reafferent Systems." *Journal of Physiology—Paris* 98 (2004): 81–97.

Carlson, Bruce A., Saad M. Hasan, Michael Hollmann, Derek B. Miller, Luke J. Harmon, and Matthew E. Arenegard. "Brain Evolution Triggers Increased Diversification of Electric Fishes." *Science* 332, no. 6029 (2011): 583–86.

Carpenter, William B. *Nature and Man: Essays Scientific and Philosophical*. London: Kegan Paul, Trench, 1888.

———. *Principles of Comparative Physiology*. 4th and rev. ed. Philadelphia: Blanchard & Lea, 1854.

Carr, Catherine E. "Neuroethology of Electric Fish." *BioScience* 40, no. 4 (1990): 259–67.

Carrubba, Robert W., and John Z. Bowers. "Engelbert Kaempfer's First Report of the Torpedo Fish of the Persian Gulf in the Late Seventeenth Century." *Journal of the History of Biology* 15, no. 2 (1982): 263–74.

Carvalho, M. R. de, L. J. V. Compagno, and P. R. Last. "Order Torpediniformes: Narcinidae: Numbfishes." In *The Living Marine Resources of the Western Central Pacific*, vol. 3: *Batoid Fishes, Chimaeras and Bony Fishes, Part 1 (Elopidae to Linophrynidae)*, edited by Kent E. Carpenter and Volker H. Niem. Rome: Food and Agriculture Organization of the United Nations, 1999.

Caton, R. "The Electric Currents of the Brain." *British Medical Journal* 2 (1875): 278.

Cavallo, Tiberius. "Of the Methods of Manifesting the Presence, and Ascertaining the Quality, of Small Quantities of Natural or Artificial Electricity." *Philosophical Transactions of the Royal Society of London* 78 (1788): 1–22.

Cavendish, Henry. "An Account of Some Attempts to Imitate the Effects of the Torpedo by Electricity." *Philosophical Transactions of the Royal Society of London* 66 (1776): 196–225.

Cavendish, Henry, and James Clerk Maxwell. *The Electrical Researches of the Honourable Henry Cavendish, F. R. S., Written between 1771 and 1781*. Cambridge: Cambridge University Press, 1879.

Chadwick, Owen. *The Secularization of the European Mind in the Nineteenth Century*. Cambridge: Cambridge University Press, 1990.

Chaisson, Eric. *Cosmic Evolution: The Rise of Complexity in Nature*. Cambridge, MA: Harvard University Press, 2001.

———. *Epic of Evolution: Seven Ages of the Cosmos*. New York: Columbia University Press, 2005.

Chalmers, Gordon Keith. "The Lodestone and the Understanding of Matter in Seventeenth Century England." *Philosophy of Science* 4, no. 1 (1937): 75–95.

Chamberlain, Bobby J., and Ronald M. Harmon. *A Dictionary of Informal Brazilian Portuguese with English Index*. Washington, DC: Georgetown University Press, 2003.

Chambers, Robert. *Vestiges of the Natural History of Creation, and Other Evolutionary Writings*. Edited by James A. Secord. Chicago: University of Chicago Press, 1994.

Chambers Journal. "Curiosities of the Telephone." *Choice Literature* 1 (1883): 245–47.

Channell, David F. *The Vital Machine: A Study of Technology and Organic Life*. Oxford: Oxford University Press, 1991.

Chapin, James. "Date: 12/28/1909 to 1/10/1910, Locality: Bafwaboka." In *American Museum Congo Expedition 1909–1915: The Diaries of James Chapin, Book 3: Nov. 1, 1909 to Feb. 5, 1910*. New York: American Museum of Natural History, 2003. http://diglib1.amnh.org/articles/chapin_diary/chapin_diary.html.

Christensen, Dan Charly. "Ørsted's *Concept of Force and Theory of Music.*" In Brain, Cohen, and Knudsen, *Hans Christian Ørsted*.

Christian, David. *Maps of Time: An Introduction to Big History*. Berkeley: University of California Press, 2004.

Cicero. *De Natura Deorum*. Translated by H. Rackham. Cambridge, MA: Harvard University Press, 1967. http://archive.org/details/denaturadeorumac oociceuoft.

Cipolla, Carlo M. *Clocks and Culture, 1300–1700*. New York: W. W. Norton, 1978.

Clark, Andy. *Being There: Putting Brain, Body, and World Together Again*. Cambridge, MA: MIT Press, 1996.

———. *Natural-Born Cyborgs: Minds, Technologies, and the Future of Human Intelligence*. Oxford: Oxford University Press, 2003.

Clarke, Edwin, and L. S. Jacyna. *Nineteenth-Century Origins of Neuroscientific Concepts*. Berkeley: University of California Press, 1987.

Clarke, Edwin, and C. D. O'Malley. *The Human Brain and Spinal Cord: A Historical Study Illustrated by Writings from Antiquity to the Twentieth Century*. 2nd ed. San Francisco: Norman Publishing, 1996.

Coates, Christopher W. "The Kick of an Electric Eel." *Atlantic Monthly* 180 (1947): 75–79.

Collins, H. M. *Changing Order: Replication and Induction in Scientific Practice*. Chicago: University of Chicago Press, 1992.

Commonwealth Bureau of Census and Statistics of Australia. *Official Year Book of the Commonwealth of Australia*. Melbourne, 1924.

Cooper, Mark S., and Manfred Schliwa. "Motility of Cultured Fish Epidermal Cells in the Presence and Absence of Direct Current Electric Fields." *Journal of Cell Biology* 102, no. 4 (1986): 1384–99.

Copenhaver, Brian P. "Did Science Have a Renaissance?" *Isis* 83, no. 3 (1992): 387–407.

———. "A Tale of Two Fishes: Magical Objects in Natural History from Antiquity through the Scientific Revolution." *Journal of the History of Ideas* 52, no. 3 (1991): 373–98.

Cotterell, Brian, and Johan Kamminga. *Mechanics of Pre-Industrial Technology: An Introduction to the Mechanics of Ancient and Traditional Material Culture*. Cambridge: Cambridge University Press, 1992.

Couch, Jonathan. *A History of the Fishes of the British Islands*. London: Groombridge, [1867].

Cox, R. T. "Electric Fish." *American Journal of Physics* 11 (1943): 13–22.

Crampton, W. G. R. "Gymnotiform Fish: An Important Component of Amazonian Floodplain Fish Communities." *Journal of Fish Biology* 48, no. 2 (1996): 298–301.

Cranefield, Paul F. "The Organic Physics of 1847 and the Biophysics of Today." *Journal of the History of Medicine and Allied Sciences* 12, no. 10 (1957): 407–23.

———. "The Philosophical and Cultural Interests of the Biophysics Movement of 1847." *Journal of the History of Medicine and Allied Sciences* 21, no. 1 (1966): 1–7.

Cressey, Daniel. "Please Sequence My Eel." *Nature News* (2008): 1p.

Cressler, John D. *Silicon Earth: Introduction to the Microelectronics and Nanotechnology Revolution*. New York: Cambridge University Press, 2009.

Crosby, Alfred W. *The Measure of Reality: Quantification and Western Society, 1250–1600*. Cambridge: Cambridge University Press, 1997.

Crosse, Andrew. "On the Production of Insects by Voltaic Electricity." *Annals of Electricity* 1, no. 46 (1837): 242–44.

[Crosse, Cornelia A. H.]. *Memorials, Scientific and Literary, of Andrew Crosse, the Electrician*. London: Longman, Brown, Green, Longmans & Roberts, 1857.

Curtis, Thomas. "Electro-Galvanism." In *London Encyclopedia*, vol. 8, edited by Thomas Curtis. London: Thomas Tegg, 1829.

Dalby, Andrew. *Food in the Ancient World, from A to Z*. London: Routledge, 2003.

Daniell, J. Frederic. "Animal Electricity." In *An Introduction to the Study of Chemical Philosophy: Being a Preparatory View of the Forces Which Concur to the Production of Chemical Phenomena*. London: John W. Parker, 1839.

Darnton, Robert. *Mesmerism and the End of the Enlightenment in France*. Cambridge, MA: Harvard University Press, 1968.

Darrigol, Olivier. "Foundations." In *Electrodynamics from Ampère to Einstein*. Oxford: Oxford University Press, 2003.

Darwin, Charles. *Charles Darwin's Beagle Diary*. Edited by R. D. Keynes. Cambridge: Cambridge University Press, 1988. http://darwin-online.org.uk/.

———. *Charles Darwin's Natural Selection: Being the Second Part of His Big Species Book Written from 1856 to 1858*. Edited by R. C. Stauffer. Cambridge: Cambridge University Press, 1975. http://darwin-online.org.uk/.

———. *The Foundations of the Origin of Species: Two Essays Written in 1842 and 1844*. Edited by F. Darwin. Cambridge: Cambridge University Press, 1909. http://darwin-online.org.uk/.

———. *Journal of Researches into the Natural History and Geology of the Countries Visited during the Voyage of HMS Beagle Round the World*. 2nd ed. London: John Murray, 1845. http://darwin-online.org.uk/.

———. *The Life and Letters of Charles Darwin, Including and Autobiographical Chapter*. Vol. 3. Edited by Francis Darwin. London: John Murray, 1887. http://darwin-online.org.uk/.

———. *More Letters of Charles Darwin*. Vol. 1. Edited by Francis Darwin. London: John Murray, 1903. http://darwin-online.org.uk/.

———. *On the Origin of Species by Means of Natural Selection, or the Preservation of Favoured Races in the Struggle for Life*. 1st ed. London: John

Murray, 1859. Fifth Thousand, 2nd ed. London: John Murray, 1860. 3rd ed. London: John Murray, 1861. 4th ed. London: John Murray, 1866. 5th ed. London: John Murray, 1869. 6th ed. London: John Murray, 1872. http://darwin-online.org.uk/.

Darwin, Erasmus. *Zoonomia; or, the Laws of Organic Life*. Vol. 1. London: J. Johnson, 1794.

Daston, Lorraine. "Objectivity and the Escape from Perspective." *Social Studies of Science* 22, no. 4 (1992): 597–618.

Daston, Lorraine, and Peter Galison. "The Image of Objectivity." *Representations* 40 (1992): 81–128.

Davidson, Iain, and William C. McGrew. "Stone Tools and the Uniqueness of Human Culture." *Journal of the Royal Anthropological Institute* 11, no. 4 (2005): 793–817.

Davy, Humphry. "An Account of Some Experiments on the Torpedo." *Philosophical Transactions of the Royal Society of London* 119 (1829): 15–18.

Davy, John. "An Account of Some Experiments and Observations on the Torpedo (Raia Torpedo, Linn.)." *Transactions of the Royal Society of London* 122 (1832): 259–78.

———. "Observations on the Torpedo, with an Account of Some Additional Experiments on Its Electricity." *Transactions of the Royal Society of London* 124 (1834): 531–50.

Dear, Peter. "From Truth to Disinterestedness in the Seventeenth Century." *Social Studies of Science* 22, no. 4 (1992): 619–31.

———. "*Totius in Verba*: Rhetoric and Authority in the Early Royal Society." *Isis* 76, no. 2 (1985): 144–61.

Debru, A. "The Power of Torpedo Fish as a Pathological Model to the Understanding of Nervous Transmission in Antiquity." *Comptes Rendus—Biologies* 329 (2006): 298–302.

DeLanda, Manuel. *Philosophy and Simulation: The Emergence of Synthetic Reason*. London: Continuum, 2011.

Delbourgo, James. *A Most Amazing Scene of Wonders: Electricity and Enlightenment in Early America*. Cambridge, MA: Harvard University Press, 2006.

Denkinger, Emma Marshall. "The Arcadia and 'The Fish Torpedo Faire.'" *Studies in Philology* 28 (1931): 162–83.

Desmond, Adrian. *The Politics of Evolution: Morphology, Medicine, and Reform in Radical London*. Chicago: University of Chicago Press, 1989.

Desmond, Adrian, and James Moore. *Darwin: The Life of a Tormented Evolutionist*. New York: W. W. Norton, 1992.

Dierig, Sven. "Engines for Experiment: Laboratory Revolution and Industrial Labor in the Nineteenth-Century City." *Osiris* 18 (2003): 116–34.

———. "Urbanization, Place of Experiment, and How the Electric Fish Was Caught by Emil Du Bois-Reymond." *Journal of the History of the Neurosciences* 9, no. 1 (2000): 5–13.

Dixon, E. James. "Human Colonization of the Americas: Timing, Technology, and Process." *Quaternary Science Reviews* 20, no. 1–3 (2001): 277–99.

Dombois, Florian. "The 'Muscle Telephone': The Undiscovered Start of Audification in the 1870s." In *Sounds of Science—Schall Im Labor (1800–1930)*, edited by Julia Kursell. Preprint 346. Berlin: Max Planck Institute for the History of Science, 2008. http://www.mpiwg-berlin.mpg.de/Preprints/P346.PDF.

Donaldson, P. E. K. "Mr. Nicholson's Doubler: An 18th-Century Programmable Charge Amplifier." *Engineering Science and Education Journal* 7, no. 2 (1998): 67–70.

Douard, John W. "E.-J. Marey's Visual Rhetoric and the Graphic Decomposition of the Body." *Studies in History and Philosophy of Science*, part A, 26, no. 2 (1995): 175–204.

Drewer, Lois. "Fisherman and Fish Pond: From the Sea of Sin to the Living Waters." *Art Bulletin* 63, no. 4 (1981): 533–47.

Driscol, Thomas E., Oscar D. Ratnoff, and Oddvar F. Nygaard. "The Remarkable Dr. Abildgaard and Countershock. The Bicentennial of His Electrical Experiments on Animals." *Annals of Internal Medicine* 83, no. 6 (1975): 878–82.

Du Bois-Reymond, Emil. "11. Observations and Experiments on Malapterurus Brought to Berlin Alive; 12 & 13. Living Torpedos in Berlin." In *Memoirs on the Physiology of Nerve, of Muscle, and of the Electrical Organ*, edited by John Scott Burdon-Sanderson. Oxford: Clarendon Press, 1887.

Dunn, P. M. "Galen (AD 129–200) of Pergamun: Anatomist and Experimental Physiologist." *Archives of Disease in Childhood: Fetal and Neonatal Edition* 88 (2003): F441–43.

Ebers, Georg. *Egypt: Descriptive, Historical, and Picturesque*, vol. 1. Translated by Clara Bell. London: Cassell, 1887.

Edgerton, David. *The Shock of the Old: Technology and Global History since 1900*. New York: Oxford University Press, 2006.

Edis, Taner. "Grand Themes, Narrow Constituency." In *Why Intelligent Design Fails: A Scientific Critique of the New Creationism*, edited by Matt Young and Taner Edis. Piscataway, NJ: Rutgers University Press, 2006.

Elliott, Paul. "Abraham Bennet, F. R. S. (1749–1799): A Provincial Electrician in Eighteenth-Century England." *Notes and Records of the Royal Society of London* 53, no. 1 (1999): 59–78.

Elsenaar, Arthur, and Remko Scha. "Electric Body Manipulation as Performance Art: A Historical Perspective." *Leonardo Music Journal* 12 (2002): 17–28.

El-Shahawy, Abeer. "The Narmer Palette." In *The Egyptian Museum in Cairo: A Walk through the Alleys of Ancient Egypt*. Photography by Farid Atiya. Cairo: American University of Cairo, 2006.

*Encylopédie méthodique. Histoire naturelle, tome troisieme: Contenant les poissons*. Paris: Chez Pancoucke Librairie, 1787.

Erasmus. *Collected Works of Erasmus: Adages III iv 1 to IV ii 100*. Edited by John N. Grant. Translated by Denis L. Drysdall. Toronto: University of Toronto Press, 2005.

Erlandson, Jon M. "The Archaeology of Aquatic Adaptations: Paradigms for a New Millennium." *Journal of Archaeological Research* 9, no. 4 (2001): 287–350.

Erlandson, Jon M., and Madonna L. Moss. "Shellfish Feeders, Carrion Eaters, and the Archaeology of Aquatic Adaptations." *American Antiquity* 66, no. 3 (2001): 413–32.

Erlandson, Jon M., and Torben C. Rick. "Archaeology Meets Marine Ecology: The Antiquity of Maritime Cultures and Human Impacts on Marine Fisheries and Ecosystems." *Annual Review of Marine Science* 2 (2010): 231–51.

Eschmeyer, William N., and Earl S. Herald. *A Field Guide to Pacific Coast Fishes*. Illustrated by Howard E. Hammann and Katherine P. Smith. Boston: Houghton Mifflin, 1983.

Evelyn, John. *The Diary of John Evelyn*. Edited by Guy la de Bédoyère. Woodbridge, UK: Boydell Press, 2004.

Fa, Darren Andrew. "Effects of Tidal Amplitude on Intertidal Resource Availability and Dispersal Pressure in Prehistoric Human Coastal Populations: The Mediterranean-Atlantic Transition." *Quaternary Science Reviews* 27, no. 23–24 (2008): 2194–209.

Fahlberg, Samuel. "Description of an Electric Eel of Surinam [from Gilbert's *Annalen Der Physik*]." *Naval Chronicle* 10 (1803): 478–80.

Fairservis, W. A. "A Revised View of the Na'rmr Palette." *Journal of the American Research Center in Egypt* 28 (1991): 1–20.

Fara, Patricia. "An Attractive Therapy: Animal Magnetism in Eighteenth-Century England." *History of Science* 33, no. 2 (1995): 127–77.

———. *An Entertainment for Angels: Electricity in the Enlightenment*. New York: Columbia University Press, 2003.

Faraday, Michael. *The Correspondence of Michael Faraday*, vol. 2: *1832–December 1840*. Edited by Frank A. J. L. James. London: Institution of Electrical Engineers, 1993.

———. "Identity of Electricity Derived from Different Sources. Relation by Measure of Common and Voltaic Electricity. Experimental Researches in Electricity. Third Series." *Philosophical Transactions of the Royal Society of London* 123 (1833): 23–54.

———. "Notice of the Character and Direction of the Electric Force of the Gymnotus. Experimental Researches in Electricity. Fifteenth Series." *Philosophical Transactions of the Royal Society of London* 129 (1839): 1–12.

Feldberg, W., and A. Fessard. "The Cholinergic Nature of the Nerves to the Electric Organ of the Torpedo (Torpedo Marmorata)." *Journal of Physiology* 101, no. 2 (1942): 200–216.

Feynman, Richard P., Robert B. Leighton, and Matthew Sands. *The Feynman Lectures on Physics*. New millennium ed. New York: Basic Books, 2010.

Finger, Stanley. "Dr. Alexander Garden, a Linnaean in Colonial America, and the Saga of Five 'Electric Eels.'" *Perspectives in Biology and Medicine* 53, no. 3 (2010): 388–406.

———. "Edgar D. Adrian: Coding in the Nervous System." In *Minds behind the Brain: A History of Pioneers and Their Discoveries*. Oxford: Oxford University Press, 2000.

———. "Edward Bancroft's 'Torporific Eels.'" *Perspectives in Biology and Medicine* 52, no. 1 (2009): 61–79.

———. "Luigi Galvani: Electricity and the Nerves." In *Minds behind the Brain.*

Finger, Stanley, and Ian Ferguson. "The Role of *The Gentleman's Magazine* in the Dissemination of Knowledge about Electric Fish in the Eighteenth Century." *Journal of the History of the Neurosciences* 18, no. 4 (2009): 347–65.

Finger, Stanley, and Mark B. Law. "Karl August Weinhold and His 'Science' in the Era of Mary Shelley's *Frankenstein:* Experiments on Electricity and the Restoration of Life." *Journal of the History of Medicine and Allied Sciences* 53, no. 2 (1998): 161–80.

Finkelstein, Gabriel. "M. Du Bois-Reymond Goes to Paris." *British Journal for the History of Science* 36, no. 130 (2003): 261–300.

Fischer, W. G., G. Bianchi, and W. B. Scott, eds. "Batoid Fishes." In *FAO Species Identification Sheets for Fishery Purposes. Eastern Central Atlantic; Fishing Areas 34, 47 (in Part)*. Ottawa: Department of Fisheries and Oceans Canada, by arrangement with the Food and Agriculture Organization of the United Nations, 1981.

Fitzsimons, J. T. "Physiology during the Nineteenth Century." *Proceedings of the Physiological Society* (1976): 16P–25P.

Flagg, Henry Collins. "Observations on the Numb Fish, or Torporific Eel." *Transactions of the American Philosophical Society* 2 (1786): 170–73.

Fleagle, John G., Zelalem Assefa, Francis H. Brown, and John J. Shea. "Paleoanthropology of the Kibish Formation, Southern Ethiopia: Introduction." *Journal of Human Evolution* 55, no. 3 (2008): 360–65.

Floreano, Dario, and Claudio Mattiussi. *Bio-Inspired Artificial Intelligence: Theories, Methods, and Technologies*. Cambridge, MA: MIT Press, 2008.

Focaccia, Miriam, and Raffaella Simili. "Luigi Galvani, Physician, Surgeon, Physicist: From Animal Electricity to Electro-Physiology." In Whitaker, Smith, and Finger, *Brain, Mind, and Medicine*.

Folgering, Joost H. A., and Bert Poolman. "Channel Electrophysiology: History, Current Applications, and Future Prospects." In *How to Escape from a Tense Situation: Bacterial Mechanosensitive Channels*. Groningen, NL: University of Groningen, 2005. http://irs.ub.rug.nl/ppn/285537504.

Fowler, Richard. *Experiments and Observations Relative to the Influence Lately Discovered by M. Galvani, and Commonly Called Animal Electricity*. Edinburgh: Printed for T. Duncan, P. Hill, Robertson & Berry, and G. Mudie; and J. Johnson, 1793.

Fraden, Jacob. *Handbook of Modern Sensors: Physics, Designs, and Applications*. 3rd ed. New York: Springer, 2003.

Frank, Robert G., Jr. "Instruments, Nerve Action, and the All-or-None Principle." *Osiris* 9 (1994): 208–35.

Franklin, Benjamin. *Animal Magnetism: Report of Dr. Franklin and Other Commissioners*. Philadelphia: H. Perkins, 1837.

———. *The Works of Benjamin Franklin*. Boston: Charles Tappan, 1844. Boston: Whittemore, Niles, & Hall, 1856.

Friedman, Michael. "Kant—Naturphilosophie—Electromagnetism." In Brain, Cohen, and Knudsen, *Hans Christian Ørsted*.

Frisch, K. von. "The Sense of Hearing in Fish." *Nature* 141 (1938): 8–11.

Fulhame, Elizabeth. *An Essay on Combustion: With a View to a New Art of Dying and Painting*. J. Cooper, 1794.

Gardner-Thorpe, Christopher, and John Pearn. "Erasmus Darwin (1731–1802): Neurologist." *Neurology* 66, no. 12 (2006): 1913–16.

Garratt, Alfred Charles. *Medical Electricity: Embracing Electro-Physiology and Electricity as a Therapeutic, with Special Reference to Practical Medicine*. Philadelphia: J. B. Lippincott, 1866.

Geddes, Leslie A. "The Capillary Electrometer: The First Graphic Recorder of Bioelectric Signals." *Archives internationales d'histoire des sciences* 14 (1961): 275–90.

———. "Did Wheatstone Build a Bridge?" *IEEE Engineering in Medicine and Biology* 25, no. 3 (2006): 88–90.

———. "What Did Caton See?" *Electroencephalography and Clinical Neurophysiology* 67, no. 1 (1987): 2–6.

Geddes, Leslie A., and L. E. Geddes. "How Did Georg Simon Ohm Do It?" *IEEE Engineering in Medicine and Biology* 17, no. 3 (1998): 107–9.

Gee, Brian. "The Early Development of the Magneto-Electric Machine." *Annals of Science* 50, no. 2 (1993): 101–33.

"General Notes: Zoology. Vertebrates." *American Naturalist* 17, no. 2 (1883): 984–85.

Geoffroy Saint-Hillaire, Étienne. "Sur l'anatomie comparée des organes électriques de la raie torpille, du gymnote engourdissant, et du silure trembleur." *Annales du Muséum National d'Histoire Naturelle* 1 (1802): 392–407.

Geribàs, Núria, Marina Mosquera, and Josep Maria Vergès. "What Novice Knappers Have to Learn to Become Expert Stone Toolmakers." *Journal of Archaeological Science* 37, no. 11 (2010): 2857–70.

Gibson, J. J. "The Theory of Affordances." In *Perceiving, Acting, and Knowing: Toward an Ecological Psychology*, edited by R. E. Shaw and J. Bransford. Hillsdale, NJ: Erlbaum, 1977.

Gilbert, William. *On the Lodestone and Magnetic Bodies, and on the Great Magnet the Earth*. Translated by P. Fleury Mottelay. New York: John Wiley and Sons, 1893.

Gill, Sydney. "A Voltaic Enigma and a Possible Solution to It." *Annals of Science* 33, no. 4 (1976): 351–70.

Gillispie, Charles Coulston. "Scientific Aspects of the French Egyptian Expedition 1798–1801." *Proceedings of the American Philosophical Society* 133, no. 4 (1989): 447–74.

Glassman, Robert B., and Hugh W. Buckingham. "David Hartley's Neural Vibrations and Psychological Associations." In Whitaker, Smith, and Finger, *Brain, Mind, and Medicine*.

Glotzhaber, Robert C., John M. Fricke, Laurence L. Meissner, and John N. Moore. *Student Laboratory Manual, Teacher's Guide for Biology: A Search for Order in Complexity*. 2nd ed. Arlington Heights, IL: Christian Liberty Press, 2005.

Golinski, Jan. *Science as Public Culture: Chemistry and Enlightenment in Britain, 1760–1820*. Cambridge: Cambridge University Press, 1992.

Golubtsov, A. S., and P. B. Berendzen. "Morphological Evidence for the Occurrence of Two Electric Catfish (Malapterurus) Species in the White Nile and Omo-Turkana Systems (East Africa)." *Journal of Fish Biology* 55, no. 3 (1999): 492–505.

Goodsell, David S. *The Machinery of Life*. 2nd ed. New York: Copernicus Books, 2009.

Goodsir, Professor. "Anatomical Details of the New Species of Malapterurus." *Proceedings of the Royal Physical Society of Edinburgh* 1 (1858): 21–22.

Gould, Stephen Jay. *Ontogeny and Phylogeny*. Cambridge, MA: Harvard University Press, 1977.

Grafton, Anthony, April Shelford, and Nancy Siraisi. *New Worlds, Ancient Texts: The Power of Tradition and the Shock of Discovery*. Cambridge, MA: Belknap Press of Harvard University Press, 1992.

Grant, Edward. *A History of Natural Philosophy: From the Ancient World to the Nineteenth Century*. Cambridge: Cambridge University Press, 2007.

Green, C. D. "Charles Babbage, the Analytical Engine, and the Possibility of a 19th-Century Cognitive Science." *History of Psychology* 8 (2005): 35–45.

Griffiths, David J. *Introduction to Electrodynamics*. 3rd ed. Upper Saddle River, NJ: Prentice Hall, 1999.

Grove, William Robert. "An Address on the Importance of the Study of Physical Science in Medical Education." *British Medical Journal* 1 (1869): 485–87.

———. *On the Correlation of Physical Forces*. London: London Institution, 1846.

Grundfest, Harry. "The Different Careers of Gustav Fritsch (1838–1927)." *Journal of the History of Medicine and Allied Sciences* 18, no. 2 (1963): 125–29.

Gudger, E. W. "The Five Great Naturalists of the Sixteenth Century: Belon, Rondelet, Salviani, Gesner, and Aldrovandi: A Chapter in the History of Ichthyology." *Isis* 22, no. 1 (1934): 21–40.

Guerrini, Anita. "Alexander Monro *Primus* and the Moral Theatre of Anatomy." *Eighteenth Century: Theory and Interpretation* 47, no. 1 (2006): 1–18.

———. *Experimenting with Humans and Animals: From Galen to Animal Rights*. Baltimore, MD: Johns Hopkins University Press, 2003.

Habgood, Philip J., and Natalie R. Franklin. "The Revolution That Didn't Arrive: A Review of the Pleistocene Sahul." *Journal of Human Evolution* 55, no. 2 (2008): 187–222.

Hackmann, W. D. *Electricity from Glass: The History of the Frictional Electrical Machine, 1600–1850*. Alphen aan den Rijn: Sijhoff & Noordhiff, 1978.

———. "The Relationship between Concept and Instrument Design in Eighteenth-Century Experimental Science." *Annals of Science* 36, no. 3 (1979): 205–24.

Hall, A. Rupert, and Marie Boas Hall. "Introduction to Part III." In *Unpublished Scientific Papers of Isaac Newton: A Selection from the Portsmouth Collection in the University Library, Cambridge*. Cambridge: Cambridge University Press, 1978.

Hall, Vance M. D. "The Contribution of the Physiologist, William Benjamin Carpenter (1813–1885), to the Development of the Principles of the Correlation of Forces and the Conservation of Energy." *Medical History* 23, no. 2 (1979): 139–55.

Haller, Albrecht von. "A Dissertation on the Sensible and Irritable Parts of Animals." *Bulletin of the History of Medicine* 4 (1936): 651–99.

———. *First Lines of Physiology*. Edited by William Cullen. Edinburgh: Charles Elliot, 1786.

Halstead, Bruce W. "Poisonous Fishes." *Public Heath Reports (1896–1970)* 73, no. 4 (1958): 302–12.

Hancock, Paul L., and Brian J. Skinner, eds. *The Oxford Companion to the Earth*. Oxford: Oxford University Press, 2000.

Hands, Joseph. "Animal Electro-Magnetism." In *New Views of Matter, Life, Motion, and Resistance: Also an Enquiry into the Materiality of Electricity, Heat, Light, Colours, and Sound*. London: E. W. Allen, 1879.

Hankins, Thomas L. *Science and the Enlightenment*. Cambridge: Cambridge University Press, 1985.

Hankins, Thomas L., and Robert J. Silverman. *Instruments and the Imagination*. Princeton, NJ: Princeton University Press, 1995.

Haring, Kristen. *Ham Radio's Technical Culture*. Cambridge, MA: MIT Press, 2007.

Harmon, L. D., and E. R. Lewis. "Neural Modeling." *Physiological Reviews* 46, no. 3 (1966): 513–91.

Harris, Lauren Julius, and Jason B. Almerigi. "Probing the Human Brain with Stimulating Electrodes: The Story of Roberts Batholow's (1874) Experiment on Mary Rafferty." *Brain and Cognition* 70, no. 1 (2009): 92–115.

Harvey, Karen. *Reading Sex in the Eighteenth Century: Bodies and Gender in English Erotic Culture*. Cambridge: Cambridge University Press, 2008.

Haüy, René Just. "Electricity." In *An Elementary Treatise on Natural Philosophy*. Translated by Olinthus Gregory. London: Printed for George Kearsley, 1807.

Hayes, Brian. *Infrastructure: A Field Guide to the Industrial Landscape*. New York: W. W. Norton, 2005.

Heilbron, John L. *Electricity in the 17th and 18th Centuries: A Study in Early Modern Physics*. Mineola, NY: Dover, 1999.

———. "Franklin, Haller, and Franklinist History." *Isis* 68, no. 4 (1977): 539–49.

Heiligenberg, Walter. *Neural Networks in Electric Fish*. Cambridge, MA: MIT Press, 1991.

Heiligenberg, Walter, and J. Bastian. "The Electric Sense of Weakly Electric Fish." *Annual Review of Physiology* 46 (1984): 561–83.

Henry, Joseph, and Anne Coleman. *The Soul of an African People: The Bambara, Their Psychic, Ethical, Religious, and Social Life*. Aschendorff: Münster i. W., 1910.

Hero of Alexandria. *The Pneumatics of Hero of Alexandria from the Original*

*Greek*. Translated by Bennet Woodcroft. London: Taylor Walton & Maberly, 1851.

Herschel, John Frederick William. *A Preliminary Discourse on the Study of Natural Philosophy*. New ed. London: Longman, Hurst, Rees, Orme, Brown, Green, 1830.

Hill, Andrew. "Disarticulation and Scattering of Mammal Skeletons." *Paleobiology* 5, no. 3 (1979): 261–74.

Hill, Andrew, and Anna K. Behrensmeyer. "Disarticulation Patterns of Some Modern East African Mammals." *Paleobiology* 10, no. 3 (1984): 366–76.

Hintz, Eric S. "Portable Power: Inventor Samuel Ruben and the Birth of Duracell." *Technology and Culture* 50, no. 1 (2009): 24–57.

Hirshfeld, Alan. *The Electric Life of Michael Faraday*. New York: Walker & Company, 2006.

*The History of Paris: From the Earliest Period to the Present Day*. London: A. & W. Galignani, 1825.

Hoff, Hebbel E., and Leslie A. Geddes. "Graphical Recording before Carl Ludwig: A Historical Summary." *Archives Internationales d'Histoire des Sciences* 12 (1959): 1–25.

———. "The Rheotome and Its Prehistory: A Study in the Historical Interrelation of Electrophysiology and Electromechanics." *Bulletin of the History of Medicine* 31, no. 3–4 (1957): 212–34, 327–47.

Hoff, Hebbel E., Leslie A. Geddes, and R. Guillemin. "The Anemograph of Ons-En-Bray: An Early Self-Registering Predecessor of the Kymograph with Translations of the Original Description and a Biography of the Inventor." *Journal of the History of Medicine and Allied Sciences* 12, no. 10 (1957): 424–48.

Hofmann, James R. *André-Marie Ampère: Enlightenment and Electrodynamics*. Cambridge: Cambridge University Press, 1995.

Holder, C. F. "Electricians of the Sea." *Frank Leslie's Popular Monthly [The American Magazine]* 21 (1886): 632–37.

Home, Roderick W. "Electricity and the Nervous Fluid." *Journal of the History of Biology* 3, no. 2 (1970): 235–51.

———. "Newton on Electricity and the Aether." In *Contemporary Newtonian Research*. Edited by Zev Bechler. Dordrecht: D. Reidel, 1982.

Hope, Adrian. "100 Years of Microphones." *New Scientist* 78, no. 1102 (1978): 378–79.

Hopkins, Carl D. "Convergent Designs for Electrogenesis and Electroreception." *Current Opinion in Neurobiology* 5 (1995): 769–77.

———. "Design Features for Electric Communication." *Journal of Experimental Biology* 202 (1999): 1217–28.

———. "Neuroethology of Electric Communication." *Annual Review of Neuroscience* 11 (1988): 497–535.

———. "Signal Evolution in Electric Communication." In *The Design of Animal Communication*, edited by Marc D. Hauser and Mark Konishi. Cambridge, MA: MIT Press, 2003.

Horapollo. *The Hieroglyphics of Horapollo*. Translated by George Boas. Princeton, NJ: Princeton University Press, 1993.

Horowitz, Paul, and Winfield Hill. *The Art of Electronics*. 2nd ed. Cambridge: Cambridge University Press, 1989.

Hughes, Howard C. *Sensory Exotica: A World beyond Human Experience*. Cambridge, MA: MIT Press, 2001.

Humboldt, Alexander von. *Personal Narrative of Travels to the Equinoctial Regions of the New Continent, during the Years 1799–1804, by Alexander De Humboldt, and Aimé Bonpland*. Translated by Helen Maria Williams. 1st ed. Philadelphia: M. Carey, 1815. 2nd ed. London: Longman, Hurst, Rees, Orme, Brown, Green, 1825.

Humboldt, Alexander von, and Joseph Louis Gay-Lussac. "Experiments on the Torpedo, by Messrs. Humboldt and Gay-Lussac. Extracted from a Letter from M. Humboldt to M. Berthollet, Dated Rome, 15 Fructid. Year 13." *Philosophical Magazine* 23 (1806): 356–60.

Hunt, Bruce J. *The Maxwellians*. Ithaca, NY: Cornell University Press, 2005.

Hunter, John. "Anatomical Observations on the Torpedo." *Philosophical Transactions (1683–1775)* 63 (1773): 481–89.

Hutchinson, Thos J. "On the Social and Domestic Traits of the African Tribes: With a Glance at Their Superstitions, Cannibalism, Etc. Etc." *Transactions of the Ethnological Society of London* 1 (1861): 327–40.

Ingenhousz, Jan. "Extract of a Letter from Dr. Jan Ingenhousz, F. R. S. to Sir John Pringle, Bart. P. R. S. Containing Some Experiments on the Torpedo, Made at Leghorn, January 1, 1773 (after Having Been Informed of Those by Mr. Walsh)." *Philosophical Transactions (1683–1775)* 65 (1775): 1–4.

Ingle, D., and D. Crews. "Vertebrate Neuroethology: Definitions and Paradigms." *Annual Review of Neuroscience* 8 (1985): 457–94.

Ingram, D. "New Experiments Concerning the Torpedo." *The Student; or, the Oxford Monthly Miscellany* 1, no. 2 (1750): 49–52.

Jacobson, Marcus. *Foundations of Neuroscience*. New York: Plenum Press, 1993.

[Jardine, W.] "[Review of] on the Origin of Species." *Edinburgh New Philosophical Journal* 11 (1860): 280–89.

Jones, Bence. *Life and Letters of Faraday*. Vol. 2. Philadelphia: J. B. Lippincott, 1870.

Jones, Clive G., John H. Lawton, and Moshe Shachack. "Positive and Negative Effects of Organisms as Physical Ecosystem Engineers." *Ecology* 78, no. 7 (1997): 1946–57.

Joordens, J. C. A., F. P. Wesselingh, J. de Vos, H. B. Vonhof, and D. Kroon. "Relevance of Aquatic Environments for Hominins: A Case Study from Trinil (Java, Indonesia)." *Journal of Human Evolution* 57, no. 6 (2009): 656–71.

Jungnickel, Christa, and Russell McCormmach. *Cavendish*. Philadelphia: American Philosophical Society, 1996.

Kalmijn, A. J. "Electric and Magnetic Field Detection in Elasmobranch Fishes." *Science* 218 (1982): 916–18.

———. "The Electric Sense of Sharks and Rays." *Journal of Experimental Biology* 55 (1971): 371–83.

———. "Electro-Perception in Sharks and Rays." *Nature* 212 (1966): 1232–33.

Kanani, Nasser. "The Parthian Battery: Electric Current 2000 Years Ago?" *Gahname: Fachzeitschrift des VINI* 7 (2004): 167–204.

Keesey, John. "How Electric Fish Became Sources of Acetylcholine Receptor." *Journal of the History of the Neurosciences* 14, no. 2 (2005): 149–64.

Keithley, Joseph F. *The Story of Electrical and Magnetic Measurements: From 500 B.C. to the 1940s*. New York: John Wiley & Sons, 1999.

Kellaway, P. "The Part Played by Electric Fish in the Early History of Bioelectricity and Electrotherapy." *Bulletin of the History of Medicine* 20 (1946): 112–37.

Kettenmann, Helmut. "Alexander Von Humboldt and the Concept of Animal Electricity." *Trends in Neurosciences* 20, no. 6 (1997): 239–42.

Keyser, Paul T. "The Purpose of the Parthian Galvanic Cells: A First-Century A.D. Electric Battery Used for Analgesia." *Journal of Near Eastern Studies* 52, no. 2 (1993): 81–98.

Kipnis, Naum. "Luigi Galvani and the Debate on Animal Electricity, 1791–1800." *Annals of Science* 44, no. 2 (1987): 107–42.

———. "Scientific Controversies in Teaching Science: The Case of Volta." *Science and Education* 10 (2001): 33–49.

Kirkland, Kyle L. "High-Tech Brains: A History of Technology-Based Analogies and Models of Nerve and Brain Function." *Perspectives in Biology and Medicine* 45, no. 2 (2002): 212–23.

Klein, Richard G., Graham Avery, Kathryn Cruz-Uribe, David Halkett, John E. Parkington, Teresa E. Steele, Thomas P. Volman, and Royden Yates. "The Ysterfontein 1 Middle Stone Age Site, South Africa, and Early Human Exploitation of Coastal Resources." *Proceedings of the National Academy of Sciences of the United States of America* 101, no. 16 (2004): 5708–15.

Kleinert, Andreas. "Volta, the German Controversy on Physics and Naturphilosophie and His Relations with Johann Wilhelm Ritter." In *Nuova Voltiana: Studies on Volta and His Times*, vol. 4, edited by Fabio Bevilacqua and Lucio Fregonese. Milan: Ulrico Hoepli, 2002.

Klotz, Irving M., and Joseph J. Katz. "Two Extraordinary Electrical Experiments." *American Scholar* 60, no. 2 (1991): 247–50.

Knellwolf, Christa, and Jane Goodall. Introduction. In *Frankenstein's Science: Experimentation and Discovery in Romantic Culture, 1780–1830*, edited by Christa Knellwolf and Jane Goodall. Surrey, UK: Ashgate, 2008.

Knight, David. *Humphry Davy: Science and Power*. Cambridge: Cambridge University Press, 1992.

———. "The Physical Sciences and the Romantic Movement." *History of Science* 9 (1970): 54–75.

Koehler, Peter J. "Neuroscience in the Work of Boerhaave and Haller." In Whitaker, Smith, and Finger, *Brain, Mind, and Medicine*.

Koehler, Peter J., Stanley Finger, and Marco Piccolino. "The 'Eels' of South America: Mid-18th-Century Dutch Contributions to the Theory of Animal Electricity." *Journal of the History of Biology* 42, no. 4 (2009): 715–63.

Kuba, Michael J., Ruth A. Byrne, and Gordon M. Burghardt. "A New Method for Studying Problem Solving and Tool Use in Stingrays (Potamotrygon Castexi)." *Animal Cognition* 13 (2010): 507–13.

Kuhn, Steven L. "Evolutionary Perspectives on Technology and Technological Change." *World Archaeology* 36, no. 4 (2004): 561–70.

Kuhn, Thomas S. "Energy Conservation as an Example of Simultaneous Discovery." In *Critical Problems in the History of Science*, edited by Marshall Clagett. Madison: University of Wisconsin Press, 1959.

Laland, Kevin N., Culum Brown, and Jens Krause. "Learning in Fishes: From Three-Second Memory to Culture." *Fish and Fisheries* 4 (2003): 199–202.

Laland, Kevin N., and William Hoppitt. "Do Animals Have Culture?" *Evolutionary Anthropology* 12, no. 3 (2003): 150–59.

Laland, K. N., F. J. Odling-Smee, and M. W. Feldman. "Cultural Niche Construction and Human Evolution." *Journal of Evolutionary Biology* 14, no. 1 (2001): 22–33.

———. "Evolutionary Consequences of Niche Construction and Their Implications for Ecology." *Proceedings of the National Academy of Sciences of the United States of America* 96, no. 18 (1999): 10242–47.

———. "Niche Construction, Biological Evolution, and Cultural Change." *Behavioral and Brain Sciences* 23 (2000): 131–75.

La Mettrie, Julien Offray de. *Man a Machine, and Man a Plant*. Translated by Richard A. Watson and Maya Rybalka. Indianapolis, IN: Hackett, 1994.

Lanska, Douglas J., and Joseph T. Lanska. "Franz Anton Mesmer and the Rise and Fall of Animal Magnetism: Dramatic Cures, Controversy, and Ultimately a Triumph for the Scientific Method." In Whitaker, Smith, and Finger, *Brain, Mind, and Medicine*.

Lanza, Clara. "Electric Fish." *Science* 2, no. 30 (1881): 26–31.

———. "An Experiment upon Electric Fish Made by Galvani." *Science* 2, no. 77 (1881): 596.

Latour, Bruno. *Science in Action*. Cambridge, MA: Harvard University Press, 1987.

Leakey, Meave G., Craig S. Feibel, Raymond L. Bernor, John M. Harris, Thure E. Cerling, Kathlyn M. Stewart, Glenn W. Storrs, Alan Walker, Lars Werdelin, and Alisa J. Winker. "Lothagam: A Record of Faunal Change in the Late Miocene of East Africa." *Journal of Vertebrate Paleontology* 16, no. 3 (1996): 556–70.

Lederer, Susan E. *Subjected to Science: Human Experimentation in America before the Second World War*. Baltimore, MD: Johns Hopkins University Press, 1995.

Lehmann, Laurent. "The Adaptive Dynamics of Niche Constructing Traits in Spatially Subdivided Populations: Evolving Posthumous Extended Phenotypes." *Evolution* 62, no. 3 (2008): 549–66.

Lenoir, Timothy. "Helmholtz and the Materialities of Communication." *Osiris* 9 (1994): 184–207.

———. "Models and Instruments in the Development of Electrophysiology, 1845–1912." *Historical Studies in the Physical and Biological Sciences* 17, no. 1 (1986): 1–54.

———. *The Strategy of Life: Teleology and Mechanics in Nineteenth-Century Biology*. Chicago: University of Chicago Press, 1982.

Léry, Jean de. *History of a Voyage to the Land of Brazil, Otherwise Called America*. Translated by Janet Whatley. Berkeley: University of California Press, 1990.

Leuchtag, H. Richard. "Animal Electricity." In *Voltage-Sensitive Ion Channels: Biophysics of Molecular Excitability*. New York: Springer, 2008.

Lewes, George Henry. "Mr. Darwin's Hypothesis." *Fortnightly Review* 4 (1868): 61–80, 492–509.

Lewin Roger, R. "Origin of Bipedalism." In *Human Evolution: An Illustrated Introduction*. 4th ed. Malden, MA: Blackwell, 1999.

Lewis, Charlton T., and Charles Short. *A Latin Dictionary*. Revised and enlarged ed. Oxford: Clarendon Press, 1879.

Lewontin, R. C. "Gene, Organism, and Environment." In *Evolution from Molecules to Men*. Edited by D. S. Bendall. Cambridge: Cambridge University Press, 1983.

Lightman, Bernard V. *Victorian Popularizers of Science: Designing Nature for New Audiences*. Chicago: University of Chicago Press, 2007.

Linari, Santi. “An Inquiry into the Electro-Chemical Properties of the Torpedo.” *Literary Gazette: A Weekly Journal of Literature, Science, and the Fine Arts* 23, no. 1151 (1839): 88–89.

Lindberg, David C. *The Beginnings of Western Science: The European Scientific Tradition in Philosophical, Religious, and Institutional Context, Prehistory to A.D. 1450*. 2nd ed. Chicago: University of Chicago Press, 2007.

Lindstrom, Jon. “The Cause of Myasthenia Gravis.” *New Scientist* 69, no. 985 (1976): 228–30.

Lissmann, Hans W. “Continuous Electrical Signals from the Tail of a Fish, *Gymnarchus Niloticus* Cuv.” *Nature* 167, no. 4240 (1951): 201–2.

———. “James Gray. 14 October 1891–14 December 1975.” *Biographical Memoirs of Fellows of the Royal Society* 24 (1978): 54–70.

———. “On the Function and Evolution of Electric Organs in Fish.” *Journal of Experimental Biology* 35 (1958): 156–91.

Lissmann, Hans W., and Ken E. Machin. “The Mechanism of Object Location in *Gymnarchus Niloticus* and Similar Fish.” *Journal of Experimental Biology* 35 (1958): 451–86.

Lloyd, G. E. R. “The Invention of Nature.” In *Methods and Problems in Greek Science: Selected Papers*. Cambridge: Cambridge University Press, 1993.

———. “The Transformations of Ancient Medicine.” *Bulletin of the History of Medicine* 66 (1992): 114–32.

Locke, Hannah Sypher, and Stanley Finger. “*Gentleman's Magazine*, the Advent of Medical Electricity, and Disorders of the Nervous System.” In Whitaker, Smith, and Finger, *Brain, Mind, and Medicine*.

Lorenz, Konrad. “The Comparative Method in Studying Innate Behaviour Patterns.” In *Physiological Mechanisms in Animal Behaviour*. Symposia of the Society for Experimental Biology 4. Cambridge: Society for Experimental Biology, 1950.

———. *The Foundations of Ethology*. Translated by Robert Warren Kickert. New York: Springer-Verlag, 1981.

Lovell, Robert. *Sive Panzoologicomineralogia, or a Compleat History of Animals and Minerals*. Oxford: Printed by Hon. Hall for Jos. Godwin, 1661.

Luisi, Pier Luigi. *The Emergence of Life: From Chemical Origins to Synthetic Biology*. Cambridge: Cambridge University Press, 2006.

Lundberg, John G., John P. Sullivan, Rocío Rodiles-Hernández, and Dean A. Hendrickson. “Discovery of African Roots for the Mesoamerican Chiapas Catfish, Lacantunia Enigmatica, Requires an Ancient Intercontinental Passage." *Proceedings of the Academy of Natural Sciences of Philadelphia* 156 (2007): 39–53.

Lunn, Francis. "Electricity." In *Encyclopedia Metropolitana, Second Division: Mixed Sciences*, vol. 2, edited by Edward Smedley. London: Baldwin & Cradock, 1830.

Lyell, Charles. *Principles of Geology, Being an Attempt to Explain the Former Changes of the Earth's Surface, by Reference to Causes Now in Operation.* Vol. 2. 1st ed. London: John Murray, 1832.

Lyman, Peter, and Hal R. Varian. "How Much Information?" Berkeley: University of California, School of Information Management and Systems, 2003. www2.sims.berkeley.edu/research/projects/how-much-info-2003/.

Lyman, R. Lee. "Archaeofaunas and Butchery Studies: A Taphonomic Perspective." *Advances in Archaeological Method and Theory* 10 (1987): 249–337.

Maceroni, Colonel Francis. "An Account of Some Remarkable Electrical Phenomena Seen in the Mediterranean, with Some Physiological Deductions." *Mechanics' Magazine, Museum, Register, Journal, and Gazette* 15 (1831): 92–96, 98–100.

Machin, Ken E., and Hans W. Lissmann. "The Mode of Operation of the Electric Receptors in *Gymnarchus Niloticus*." *Journal of Experimental Biology* 37, no. 4 (1960): 801–11.

Madea, B., and C. Henssge. "Timing of Death." In *Forensic Medicine: Clinical and Pathological Aspects*, edited by Jason Payne-James, Anthony Busuttil, and William Smock. London: Greenwich Medical Media, 2003.

Mahon, Basil. *The Man Who Changed Everything: The Life of James Clerk Maxwell.* Chichester, West Sussex: John Wiley & Sons, 2004.

Malek, Jaromir. "The Old Kingdom (c. 2686–2160 BC)." In *The Oxford History of Ancient Egypt*, edited by Ian Shaw. Oxford: Oxford University Press, 2000.

Malmivuo, Jaakko, and Robert Plonsey. "A Short History of Bioelectromagnetism." In *Bioelectromagnetism: Principles and Applications of Bioelectric and Biomagnetic Fields*. New York: Oxford University Press, 1995.

Marean, Curtis W. "When the Sea Saved Humanity." *Scientific American* 303, no. 2 (2010): n.p.

Marean, Curtis W., Miryam Bar-Matthews, Jocelyn Bernatchez, Erich Fisher, Paul Goldbert, Andy I. R. Herries, Zenobia Jacobs, Antonieta Jerardino, Panagiotis Karkanas, and Tom Minichillo. "Early Human Use of Marine Resources and Pigment in South Africa during the Middle Pleistocene." *Nature* 449 (2007): 905–8.

Marey, E. J. *Animal Mechanism: A Treatise on Terrestrial and Aerial Locomotion.* London: Kegan Paul, Trench, 1883.

Markham, Clements R., ed. "A New Discovery of the Great River of the Amazons, by Father Cristoval De Acuña, a Priest of the Company of Jesus . . . in the Year 1639." In *Expeditions into the Valley of the Amazons, 1539, 1540, 1639*. London: Hakluyt Society, 1859.

Martins, Roberto de Andrade. "Ørsted, Ritter, and Magnetochemistry." In Brain, Cohen, and Knudsen, *Hans Christian Ørsted*.

Matteucci, Carlo. "Electro-Physiological Researches, First Memoir: The Muscular Current." *Philosophical Transactions of the Royal Society of London* 135 (1845): 283–96.

———. "Electro-Physiological Researches, Second Memoir: On the Proper Current of the Frog." *Philosophical Transactions of the Royal Society of London* 135 (1845): 297–302.

———. "Electro-Physiological Researches, Third Memoir: On Induced Contractions." *Philosophical Transactions of the Royal Society of London* 135 (1845): 303–18.

———. "Electro-Physiological Researches, Fourth Memoir: The Physiological Action of the Electric Current." *Philosophical Transactions of the Royal Society of London* 135 (1846): 483–99.

———. "Electro-Physiological Researches, Sixth Series: Laws of the Electric Discharge of the Torpedo and Other Electric Fishes—Theory of the Production of Electricity in These Animals." *Philosophical Transactions of the Royal Society of London* 137 (1847): 239–41.

Mauro, Alexander. "The Role of the Voltaic Pile in the Galvani-Volta Controversy Concerning Animal vs. Metallic Electricity." *Journal of the History of Medicine and Allied Sciences* 14, no. 2 (1969): 140–50.

Mazis, Glen A. *Humans, Animals, Machines: Blurring Boundaries*. Albany, NY: SUNY Press, 2008.

McBrearty, Sally, and Alison S. Brooks. "The Revolution That Wasn't: A New Interpretation of the Origin of Modern Human Behavior." *Journal of Human Evolution* 39, no. 5 (2000): 453–563.

McElroy, William D., and Bentley Glass, eds. *A Symposium on Light and Life*. Baltimore, MD: Johns Hopkins University Press, 1960.

McNeill, John R., and William McNeill. *The Human Web: A Bird's-Eye View of World History*. New York: W. W. Norton, 2003.

McNeill, William H. "History and the Scientific Worldview." In *The Return of Science: Evolution, History, and Theory*. Edited by Philip Pomper and David Gary Shaw. Lanham, MD: Rowman & Littlefield, 2002.

M'Donnell, Robert. "On an Organ in the Skate Which Appears to Be the Homologue of the Electrical Organ of the Torpedo." *Natural History Review* 1 (1861): 57–60.

———. "On the Organs Which in the Common Ray Are Homologous with the Electrical Organs of the Torpedo." *Proceedings of the Royal Irish Academy (1836–1869)* 7 (1857): 362–65.

Mellor, Anne K. "Making a 'Monster': An Introduction to Frankenstein." In *The Cambridge Companion to Mary Shelley*, edited by Esther H. Schor. Cambridge: Cambridge University Press, 2003.

Melton, J. Gordon, ed. *Encyclopedia of Occultism & Parapsychology*. 5th ed. Detroit: Gale Research, 2001.

Mercader, Julio, Melissa Panger, and Christophe Boesch. "Excavation of a Chimpanzee Stone Tool Site in the African Rainforest." *Science* 296 (2002): 1452–55.

Mertens, Joost. "Shocks and Sparks: The Voltaic Pile as a Demonstration Device." *Isis* 89, no. 2 (1998): 300–11.

Meyer, Hermann J. *Meyers Konversations-Lexicon: Eine Encyklopädie des allgemeinen Wissens*. Leipzig: Verlag des Bibliographischen Instituts, 1890.

Migeod, Frederick William Hugh. *Mende Natural History Vocabulary*. London: Kegan Paul, Trench, 1913.

Millingen, J. G. *Curiosities of Medical Experience*. 2nd rev. ed. London: Richard Bentley, 1839.

Mills, Allan A. "Early Voltaic Batteries: An Evaluation in Modern Units and Application to the Work of Davy and Faraday." *Annals of Science* 60, no. 4 (2003): 373–98.

Mims, Forrest M., III. *Electronic Sensor Circuits and Projects*. Engineer's Mini-Notebook Series 3. Lincolnwood, IL: Master Publishing, 2000.

———. *Science and Communication Circuits and Projects*. Engineer's Mini-Notebook Series 2. Lincolnwood, IL: Master Publishing, 2000.

Mindell, David A. *Between Human and Machine: Feedback, Control, and Computing before Cybernetics*. Baltimore, MD: Johns Hopkins University Press, 2004.

Mithen, Steven J. "The Evolution of Imagination: An Archaeological Perspective." *SubStance* 30, no. 1–2 (2001): 28–54.

Moller, Peter. *Electric Fishes: History and Behavior*. London: Chapman & Hall, 1995.

Montaigne, Michel de. "Apology for Raimond De Sebonde." In *Essays of Montaigne*, vol. 2, edited by William Carew Hazlitt; translated by Charles Cotton. London: Reeves & Turner, 1877.

Moore, Gordon E. "Cramming More Components onto Integrated Circuits." *Electronics* 38, no. 8 (1965): 114–17.

"More Marine Stores: The Narke." *Fraser's Magazine for Town and Country* 45 (1852): 631–35.

Morgan, James P. "The First Reported Case of Electrical Stimulation of the Brain." *Journal of the History of Medicine and Allied Sciences* 37, no. 1 (1982): 51–64.

Morus, Iwan Rhys. "Currents from the Underworld: Electricity and the Technology of Display in Early Victorian England." *Isis* 84, no. 1 (1993): 50–69.

———. "The Electric Ariel: Telegraphy and Commercial Culture in Early Victorian England." *Victorian Studies* 39, no. 3 (1996): 339–78.

———. *Frankenstein's Children: Electricity, Exhibition, and Experiment in Early-Nineteenth-Century London*. Princeton, NJ: Princeton University Press, 1998.

———. "Galvanic Cultures: Electricity and Life in the Early 19th Century." *Endeavour* 22, no. 1 (1998): 7–11.

———. "A Grand and Universal Panacea: Death, Resurrection, and the Electric Chair." In *Bodies/Machines*, edited by Iwan Rhys Morus. Oxford: Berg, 2002.

———. "Manufacturing Nature: Science, Technology, and Victorian Consumer Culture." *British Journal for the History of Science* 29, no. 4 (1996): 403–34.

———. "'The Nervous System of Britain': Space, Time, and the Electric Telegraph in the Victorian Age." *British Journal for the History of Science* 33, no. 4 (2000): 455–75.

———. "Radicals, Romantics, and Electrical Showmen: Placing Galvanism at the End of the English Enlightenment." *Notes and Records of the Royal Society of London* 63, no. 3 (2009): 263–75.

———. *When Physics Became King*. Chicago: University of Chicago Press, 2005.

Moruzzi, Giuseppe. "The Electrophysiological Work of Carlo Matteucci." *Brain Research Bulletin* 40, no. 2 (1996): 69–91.

Morwood, M. J., P. B. O'Sullivan, F. Aziz, and A. Raza. "Fission-Track Ages of Stone Tools and Fossils on the East Indonesian Island of Flores." *Nature* 392 (1998): 173–76.

Mosini, Valeria. "When Chemistry Entered the Pile." In *Nuova Voltiana: Studies on Volta and His Times*, vol. 5, edited by Fabio Bevilacqua and Lucio Fregonese. Milan: Ulrico Hoepli, 2003.

Mottelay, Paul Fleury. *Bibliographical History of Electricity and Magnetism, Chronologically Arranged. Researches Into the Domain of the Early Sciences, Especially from the Period of the Revival of Scholasticism, with Biographical and Other Accounts of the Most Distinguished Natural Philosophers Throughout the Middle Ages*. London: C. Griffith & Co., 1922.

Müller, J. "Of the Phenomena, or Active Properties, Common to Inorganic and Organic Bodies." In *Elements of Physiology*, 2nd ed., edited by John Bell; translated by William Baly. Philadelphia: Lea & Blanchard, 1843.

Murray, Andrew. "On Electrical Fishes; with a Description of a New Species of Malapterurus from Old Calabar, Received from the Rev. Hope M. Waddell, Missionary There." *Proceedings of the Royal Physical Society of Edinburgh* 1 (1858): 20–21.

———. "Remarks on Some Plants and Their Natural History Contributions from Old Calabar." *Edinburgh New Philosophical Journal* 10 (1859): 159–61.

———. "Supplemental Observations on Electric Fishes." *Edinburgh New Philosophical Journal* 2 (1855): 379.

Myers, George S. "A Brief Sketch of the History of Ichthyology in America to the Year 1850." *Copeia* 1964, no. 1 (1964): 33–41.

"Natural History. Electric Eel." *Spirit of the English Magazines* 10, no. 1 (1821): 245.

Newman, Paul. *A Hausa-English Dictionary*. New Haven, CT: Yale University Press, 2007.

Newton, Isaac. *Sir Isaac Newton's Mathematical Principles of Natural Philosophy and His System of the World*. Edited by Florian Cajori. Translated by Andrew Motte. Berkeley: University of California Press, 1934.

Nicholson, William. "A Description of an Instrument Which, by the Turning of a Winch, Produces the Two States of Electricity without Friction or Communication with the Earth." *Philosophical Transactions of the Royal Society of London* 78 (1788): 403–7.

———. "Observations on the Electrophore, Tending to Explain the Means by Which the Torpedo and Other Fish Communicate the Electric Shock." *Journal of Natural Philosophy, Chemistry, and the Arts* 1 (1797): 355–59.

Northcutt, R. Glenn. "Evolution of the Nervous System: Changing Views of Brain Evolution." *Brain Research Bulletin* 55, no. 6 (2001): 663–74.

Nysten, Pierre-Hubert. *Nouvelles experiences galvaniques, faites sur les organes musculaires de l'homme et des animaux à sang rouge*. Paris: Chez Levrault Frères, 1803.

Odling-Smee, F. John, Kevin N. Laland, and Marcus W. Feldman. "Niche Construction." *American Naturalist* 147, no. 4 (1996): 641–48.

———. *Niche Construction: The Neglected Process in Evolution*. Princeton, NJ: Princeton University Press, 2003.

Oersted, Hans Christian. "Magnetism from the Electric Current." *Science News-Letter* 21, no. 567 (1932): 118–20.

Olmstead, John W. "The Scientific Expedition of Jean Richer to Cayenne (1672–1673)." *Isis* 34 (1942): 117–28.

Olshanckiy, Vladimir. "Body-Size Electric Eye." НАУКА И ЖИЗНЬ [Science and Life, Moscow] 11 (2005). http://www.nkj.ru/en/archive/articles/2444.

Otero, Olga, Aurélie Pinton, Hassan Taisso Mackaye, Andossa Likius, Patrick

Vignaud, and Michel Brunet. "First Description of a Pliocene Ichthyofauna from Central Africa (Site KL2, Kolle Area, Eastern Djurab, Chad): What Do We Learn?" *Journal of African Earth Sciences* 54, no. 3–4 (2009): 62–74.

Otis, Laura. "Howled out of the Country: Wilkie Collins and H. G. Wells Retry David Ferrier." In *Neurology and Literature, 1860–1920*, edited by Anne Stiles. Houndmills, Basingstoke, UK: Palgrave Macmillan, 2007.

———. "The Metaphoric Circuit: Organic and Technological Communication in the Nineteenth Century." *Journal of the History of Ideas* 63, no. 1 (2002): 105–28.

———. *Müller's Lab*. Oxford: Oxford University Press, 2007.

———. *Networking: Communicating with Bodies and Machines in the Nineteenth Century*. Ann Arbor: University of Michigan Press, 2001.

Owen, Richard. "Lecture 8: Nervous System of Fishes." In *Lectures on the Comparative Anatomy and Physiology of the Vertebrate Animals, Delivered at the Royal College of Surgeons of England, in 1844 and 1846, Part 1: Fishes*. London: Longman, Brown, Green, & Longmans, 1846.

Page, Frederick G. "Henry Hill Hickman: A Shropshire Medical Practitioner." *Medical Historian: The Bulletin of the Liverpool Medical History Society* 9 (1997): 35–42.

Paley, William. *Natural Theology: Or, Evidences of the Existence and Attributes of the Deity, Collected from the Appearances of Nature*. 2nd ed. London: R. Faulder, 1802.

Pancaldi, Giuliano. "On Hybrid Objects and Their Trajectories: Beddoes, Davy, and the Battery." *Notes and Records of the Royal Society of London* 63, no. 3 (2009): 247–62.

———. *Volta: Science and Culture in the Age of Enlightenment*. Princeton, NJ: Princeton University Press, 2003.

Parent, André. "Giovanni Aldini: From Animal Electricity to Human Brain Stimulation." *Canadian Journal of Neurological Sciences* 31 (2004): 576–84.

Pattie, Frank A. "Mesmer's Medical Dissertation and Its Debt to Mead's *De Imperio Solis ac Lunae*." *Journal of the History of Medicine and Allied Sciences* 11, no. 3 (1956): 275–87.

Pauly, Daniel. *Darwin's Fishes: An Encyclopedia of Ichthyology, Ecology, and Evolution*. Cambridge: Cambridge University Press, 2004.

Pedlar, Valerie. "Experimentation or Exploitation? The Investigations of David Ferrier, Dr Benjula, and Dr Seward." *Interdisciplinary Science Reviews* 28, no. 3 (2003): 169–74.

Peters, Robert C. "Academic Freedom and the Discovery of Electroreception in Catfish and Dogfish." *Animal Biology* 58, no. 3 (2008): 275–83.

Pettigrew, John D. "Electroreception in Monotremes." *Journal of Experimental Biology* 202 (1999): 1447–54.

Piccolino, Marco. "Animal Electricity and the Birth of Electrophysiology: The Legacy of Luigi Galvani." *History of Neuroscience* 46, no. 5 (1998): 381–407.

———. "Luigi Galvani and Animal Electricity: Two Centuries after the Foundation of Electrophysiology." *Trends in Neurosciences* 20, no. 10 (1997): 443–48.

———. "The Taming of the Electric Ray: From a Wonderful and Dreadful 'Art' to 'Animal Electricity' and 'Electric Battery.'" In Whitaker, Smith, and Finger, *Brain, Mind, and Medicine.*

———. *The Taming of the Ray: Electric Fish Research in the Enlightenment, from Walsh to Volta*. Manuscript. Ferrara, Italy, 2007.

Piccolino, Marco, and Marco Bresadola. "Drawing a Spark from Darkness: John Walsh and Electric Fish." *Trends in Neurosciences* 25, no. 1 (2002): 51–57.

Pickering, Andrew. *The Cybernetic Brain: Sketches of Another Future*. Chicago: University of Chicago Press, 2009.

Plato. *Meno*. Translated by Benjamin Jowett. New York: Bigelow, Brown, 1914.

Platt, J. R. "Amplification Aspects of Biological Response and Mental Activity." *American Scientist* 44 (1956): 180–97.

Pliny the Elder. *The Natural History*. Edited by John Bostock and H. T. Riley. London: Taylor & Francis, 1855.

Plumb, Christopher. "The 'Electric Stroke' and the 'Electric Spark': Anatomists and Eroticism at George Baker's Electric Eel Exhibition in 1776 and 1777." *Endeavour* 34, no. 3 (2010): 87–94.

Plummer, Thomas. "Flaked Stones and Old Bones: Biological and Cultural Evolution at the Dawn of Technology." *American Journal of Physical Anthropology* 125 (2004): 118–64.

Pocock, R. F. "Andrew Crosse: Early Nineteenth-Century Amateur of Electrical Science." *Science, Measurement and Technology, IEE Proceedings A* 140, no. 3 (1993): 187–96.

Pontzer, Herman, David A. Raichlen, and Michael D. Sockol. "The Metabolic Cost of Walking in Humans, Chimpanzees, and Early Hominins." *Journal of Human Evolution* 56, no. 1 (2009): 43–54.

Pouillet, C. S. M. "Note on a Method of Measuring Extremely Short Intervals of Time, Such as the Duration of the Impact of Elastic Bodies, That of the Escape of Springs, the Ignition of Powder, etc., and on a New Means of Comparing the Intensities of Electric Currents, Whether Permanent or Instantaneous." *Walker's Electrical Magazine* 1 (1845): 565–70.

Priestley, Joseph. "An Account of a New Electrometer, Contrived by Mr. William Henly, and of Several Electrical Experiments Made by Him." *Philosophical Transactions (1683–1775)* 62 (1772): 359–64.

———. *The History and Present State of Electricity: With Original Experiments.* 3rd corrected revised ed. London: Printed for C. Bathhurst et al., 1775.

Purchas, Samuel. *Hakluytus Posthumus, or Purchas His Pilgrimes.* Vols. 9, 16. Glasgow: James MacLehose & Sons, 1905; 1906.

Purrington, Robert D. "Electromagnetism." In *Physics in the Nineteenth Century.* New Brunswick, NJ: Rutgers University Press, 1997.

Radcliffe, William. *Fishing from the Earliest Times.* London: John Murray, 1921.

Reeves, C., and D. Taylor. "A History of the Optic Nerve and Its Diseases." *Eye* 18 (2004): 1096–109.

Reynolds, Edward H. "Todd, Faraday, and the Electrical Basis of Brain Activity." *Lancet Neurology* 3, no. 9 (2004): 557–63.

Richards, Robert J. *The Romantic Conception of Life: Science and Philosophy in the Age of Goethe.* Chicago: University of Chicago Press, 2002.

Richardson, Benjamin Ward. "The Croonian Lecture: On Muscular Irritability after Systemic Death [Abstract]." *Proceedings of the Royal Society of London* 21 (1872): 339–48.

———. "Dr. B. W. Richardson's Lettsomian Lectures on Certain of the Phenomena of Life." *British Medical Journal* 1, no. 10 (1861): 254–58.

Riordan, Michael, and Lillian Hoddeson. *Crystal Fire: The Invention of the Transistor and the Birth of the Information Age.* New York: W. W. Norton, 1997.

Riskin, Jessica. *Science in the Age of Sensibility: The Sentimental Empiricists of the French Enlightenment.* Chicago: University of Chicago Press, 2002.

Ritter, Johann Wilhelm. *Key Texts of Johann Wilhelm Ritter (1776–1810) on the Science and Art of Nature.* Edited by Jocelyn Holland. Translated by Jocelyn Holland. Leiden: Koninklikje Brill NV, 2010.

Ritvo, Harriet. *The Platypus and the Mermaid, and Other Figments of the Classifying Imagination.* Cambridge, MA: Harvard University Press, 1997.

———. "Toward a More Peaceable Kingdom." In *Noble Cows and Hybrid Zebras: Essays on Animals and History.* Charlottesville: University of Virginia Press, 2010.

Roberts, Lissa. "Science Becomes Electric: Dutch Interaction with the Electrical Machine during the Eighteenth Century." *Isis* 90, no. 4 (1999): 680–714.

Roberts, Tyson R. "A Review of the African Electric Catfish Family *Malapteruridae*, with Descriptions of New Species." *Occasional Papers in Ichthyology* 1 (2000): 1–15.

Robinson, Charles Henry. *Dictionary of the Hausa Language*, vol. 1: *Hausa-English.* Cambridge: University of Cambridge Press, 1899.

Robinson, Joseph D. *Mechanisms of Synaptic Transmission: Bridging the Gaps (1890–1990)*. New York: Oxford University Press, 2001.

Rocca, Julius. "William Cullen (1710–1790) and Robert Whytt (1714–1766) on the Nervous System." In Whitaker, Smith, and Finger, *Brain, Mind, and Medicine*.

Romanes, George John. *The Life and Letters of George John Romanes*. London: Longmans, Green, 1896.

Ronayne, Thomas. "A Letter from Thomas Ronayne, Esq.; to Benjamin Franklin, LL.D. F.R.S. Inclosing an Account of Some Observations on Atmospheric Electricity; in Regard of Fogs, Mists, &C. with Some Remarks." *Philosophical Transactions (1683–1775)* 62 (1772): 137–46.

Rose, Gary J. "Insights into Neural Mechanisms and Evolution of Behaviour from Electric Fish." *Nature Reviews Neuroscience* 5, no. 12 (2004): 943–51.

Rossi, Paolo. *The Birth of Modern Science*. Oxford: Blackwell, 2001.

Russell, E. S. *Form and Function: A Contribution to the History of Animal Morphology*. London: John Murray, 1916.

Russell, Edmund. *Evolutionary History: Uniting History and Biology to Understand Life on Earth*. Cambridge: Cambridge University Press, 2011.

Sachs, Jessica Snyder. *Corpse: Nature, Forensics, and the Struggle to Pinpoint the Time of Death*. Cambridge, MA: Perseus Books, 2004.

Sagua, V. O. "Observations on the Food and Feeding Habits of the African Electric Catfish *Malapterurus Electricus* (Gmelin)." *Journal of Fish Biology* 15 (1979): 61–69.

Sanford, Fernando. "Contact Electrification and the Electric Current." *Scientific Monthly* 1, no. 2 (1915): 124–31.

———. "The Discovery of Contact Electrification." *Popular Science Monthly* 83 (1913): 441–49.

Sarpeshkar, Rahul. *Ultra Low Power Bioelectronics: Fundamentals, Biomedical Applications, and Bio-Inspired Systems*. Cambridge: Cambridge University Press, 2010.

Schaffer, Simon. "Exactly Like a Stingray." *London Review of Books* (June 3, 2004).

———. "Fish and Ships: Models in the Age of Reason." In *Models: The Third Dimension of Science*, edited by Soraya de Chadarevian and Nick Hopwood. Stanford, CA: Stanford University Press, 2004.

———. "Natural Philosophy and Public Spectacle in the Eighteenth Century." *History of Science* 21 (1983): 1–43.

———. "Self Evidence." *Critical Inquiry* 18, no. 2 (1992): 327–62.

Schechter, David Charles. "Early Experiences with Resuscitation by Means of Electricity." *Surgery* 69 (1971): 360–72.

———. "Origins of Electrotherapy, Part 1." *New York State Journal of Medicine* 71, no. 2 (1971): 997–1008.

———. "Origins of Electrotherapy, Part 2." *New York State Journal of Medicine* 71, no. 2 (1971): 1114–24.

Schiller, Francis. "Reverend Wesley, Doctor Marat, and Their Electric Fire." *Clio Medica* 15, no. 3–4 (1981): 159–76.

Schuetze, Stephen M. "The Discovery of the Action Potential." *Trends in Neurosciences* 6 (1983): 164–68.

Schwarz, K. K. "Faraday and Babbage." *Notes and Records of the Royal Society of London* 56, no. 3 (2002): 367–81.

Secord, James. "Extraordinary Experiment: Electricity and the Creation of Life in Victorian England." In *The Uses of Experiment: Studies in the Natural Sciences*, edited by David Gooding, Trevor Pinch, and Simon Schaffer. Cambridge: Cambridge University Press, 1989.

———. Introduction. In Robert Chambers, *Vestiges of the Natural History of Creation, and Other Evolutionary Writings*, edited by James A. Secord. Chicago: University of Chicago Press, 1994.

———. *Victorian Sensation: The Extraordinary Publication, Reception, and Secret Authorship of Vestiges of the Natural History of Creation*. Chicago: University of Chicago Press, 2000.

Seitz, Frederick, and Norman G. Einspruch. *Electronic Genie: The Tangled History of Silicon*. Urbana: University of Illinois, 1998.

Sept, Jeanne. "Archaeological Evidence and Ecological Perspectives for Reconstructing Early Hominid Subsistence Behavior." *Archaeological Method and Theory* 4 (1992): 1–56.

Serena, Fabrizio. *Field Identification Guide to the Sharks and Rays of the Mediterranean and Black Sea*. Rome: Food and Agriculture Organization of the United Nations, 2005.

Seyfarth, Ernst-August. "Julius Bernstein (1839–1917): Pioneer Neurobiologist and Biophysicist." *Biological Cybernetics* 94 (2006): 2–8.

Sharples, R. W. *Theophrastus of Eresus: Sources for His Life, Writings, Thought and Influence*, vol. 5: *Sources on Biology*. Leiden: Brill, 1995.

Shasha, Dennis, and Cathy Lazere. *Natural Computing: DNA, Quantum Bits, and the Future of Smart Machines*. New York: W. W. Norton, 2010.

Shelley, Mary Wollstonecraft. *Frankenstein; or, the Modern Prometheus*. New ed. 2 vols. London: G. & W. B. Whittaker, 1823.

———. "Original Correspondence: *Frankenstein*." *London Literary Gazette* 770 (1831): 682–83.

Shubin, Neil. *Your Inner Fish: A Journey into the 3.5-Billion-Year History of the Human Body*. New York: Vintage Books, 2008.

Simpson, J. Y. "Local Anaesthesia: Notes on Its Artificial Production by Chloroform, &c., in the Lower Animals, and in Man." *Provincial Medical and Surgical Journal (1844–1852)* 12, no. 14 (1848): 365–71.

Sleigh, Charlotte. "Life, Death, and Galvanism." *Studies in History and Philosophy of Biological and Biomedical Sciences* 29, no. 2 (1998): 219–48.

Smail, Daniel Lord. *On Deep History and the Brain*. Berkeley: University of California Press, 2007.

Smith, Bruce D. "The Ultimate Ecosystem Engineers." *Science* 315, no. 5820 (2007): 1797–98.

Smith, C. U. M. "Brain and Mind in the 'Long' Eighteenth Century." In Whitaker, Smith, and Finger, *Brain, Mind, and Medicine*.

South Kensington Museum. *Catalogue of the Special Loan Collection of Scientific Apparatus at the South Kensington Museum*. 2nd ed. London: Printed by George E. Eyre & William Spottiswoode for Her Majesty's Stationery Office, 1876.

Stark, James. "On the Existence of an Electrical Apparatus in the Flapper Skate and Other Rays." *Proceedings of the Royal Society of Edinburgh* 2, no. 25.

Stauffer, Robert C. "Persistent Errors Regarding Oersted's Discovery of Electromagnetism." *Isis* 44, no. 4 (1953): 307–10.

———. "Speculation and Experiment in the Background of Oersted's Discovery of Electromagnetism." *Isis* 48, no. 1 (1957): 33–50.

Steele, Teresa E. "A Unique Hominin Menu Dated to 1.95 Million Years Ago." *Proceedings of the National Academy of Sciences of the United States of America* 107, no. 24 (2010): 10771–72.

Sterelny, Kim. *Thought in a Hostile World: The Evolution of Human Cognition*. Malden, MA: Blackwell, 2003.

Stevens, Richard G. "Light-at-Night, Circadian Disruption and Breast Cancer: Assessment of Existing Evidence." *International Journal of Epidemiology* 38, no. 4 (2009): 963–70.

Stevenson, Lloyd G. "Suspended Animation and the History of Anesthesia." *Bulletin of the History of Medicine* 49, no. 4 (1975): 482–511.

Stewart, Kathlyn M. "Early Hominid Utilisation of Fish Resources and Implications for Seasonality and Behaviour." *Journal of Human Evolution* 27, no. 1–3 (1994): 229–45.

———. "Fossil Fish from the Nile River and Its Southern Basins." In *The Nile: Origin, Environments, Limnology, and Human Use*, edited by H. J. Dumont. Dordrecht, NL: Springer, 2009.

———. "The Freshwater Fish of Neogene Africa (Miocene–Pleistocene): Systematics and Biogeography." *Fish and Fisheries* 2, no. 3 (2001): 177–230.

Stewart, Kathlyn M., and Alison M. Murray. "Fish Remains from the Plio-Pleistocene Shungura Formation, Omo River Basin, Ethiopia." *Geobios* 41, no. 2 (2008): 283–95.

Stillings, Dennis. "Mediterranean Origins of Electrotherapy." *Journal of Bioelectricity* 2, no. 2–3 (1983): 181–86.

———. "The Piscean Origin of Medical Electricity." *Medical Instrumentation* 7 (1973): 163–64.

Strickland, Stuart. "Galvanic Disciplines: The Boundaries, Objects, and Identities of Experimental Science in the Era of Romanticism." *History of Science* 33, no. 4 (1995): 449–68.

Sudduth, William M. "Eighteeth-Century Identifications of Phlogiston with Electricity." *Ambix* 25, no. 2 (1978): 131–47.

———. "The Voltaic Pile and Electro-Chemical Theory in 1800." *Ambix* 27, no. 1 (1980): 26–35.

Sullivan, John P., John G. Lundberg, and Michael Hardman. "A Phylogenetic Analysis of the Major Groups of Catfishes (*Teleostei: Siluriformes*) Using rag1 and rag2 Nuclear Gene Sequences." *Molecular Phylogenetics and Evolution* 41 (2006): 636–62.

Surr, Thomas Skinner. *A Winter in London; or, Sketches of Fashion*. Vol. 2. 9th ed. London: Richard Phillips, 1807.

Tattersall, Ian. *The World from Beginnings to 4000 BCE*. Oxford: Oxford University Press, 2008.

Taylor, W. K. "Computers and the Nervous System." *Symposium of the Society for Experimental Biology* 14 (1960): 152–68.

*30,000 Years of Art: The Story of Human Creativity across Time and Space*. London: Phaidon, 2007.

Thomas, Brian. "Is There Evolution in the Congo River?" Institute for Creation Research, Nov. 24, 2009. http://www.icr.org/article/there-evolution-congo-river/.

Thompson, D'Arcy Wentworth. "On Egyptian Fish-Names Used by Greek Writers." *Journal of Egyptian Archaeology* 14, no. 1–2 (1928): 22–33.

Thorndyke, Lynn. "Mediaeval Magic and Science in the Seventeenth Century." *Speculum* 28, no. 4 (1953): 101–11.

Timbs, John. "The Electric Eel, at the Royal Gallery, Adelaide-Street, Strand." *Literary World: A Journal of Popular Information and Entertainment* 3, no. 56 (1840): 33–36.

———. "Electric Fishes, Their History and Application." In *Strange Stories of the Animal World: A Book of Curious Contributions to Natural History*. London: Griffith & Farran, 1866.

Todd, Janet M. *The Secret Life of Aphra Behn*. New Brunswick, NJ: Rutgers University Press, 1997.

Todd, John T. "Account of Some Experiments on the Torpedo Electricus, at La Rochelle." *Philosophical Transactions of the Royal Society of London* 107 (1817): 32–35.

———. "Some Observations and Experiments Made on the Torpedo of the Cape of Good Hope in the Year 1812." *Philosophical Transactions of the Royal Society of London* 106 (1816): 120–26.

"Traité des signes de la mort, et des moyens de prévenir les enterrements prématurés, by E. Bouchut [précis]." *Dublin Quarterly Journal of Medical Science* 12, no. 23 (1851): 107–33.

Trapani, Josh. "Quaternary Fossil Fish from the Kibish Formation, Omo Valley, Ethiopia." *Journal of Human Evolution* 55, no. 3 (2008): 521–30.

Turner, Robert. *Electricology; or, a Discourse upon Electricity. Being an Enquiry into the Nature; Causes; Properties; and Effects Thereof, upon the Principles of the Aether. Illustrated by a Series of Surprizing Experiments.* Worcester: Printed for the Author by Thomas Olivers, 1746.

Uexküll, Jakob von. *A Foray into the Worlds of Animals and Humans: With a Theory of Meaning.* Translated by Joseph D O'Neil. Minneapolis: University of Minnesota Press, 2010.

United States Geological Survey, Washington Water Science Center. "Rain: A Water Resource." http://mo.water.usgs.gov/outreach/rain/index.htm.

Ure, Andrew. "Galvanism." In *A Dictionary of Chemistry.* 3rd ed. London: Thomas Tegg, 1827.

van der Eijk, Philip. "The Role of Medicine in the Formation of Early Greek Thought." In *The Oxford Handbook of Presocratic Philosophy*, edited by Patricia Curd and Daniel W. Graham. Oxford: Oxford University Press, 2008.

van Wyhe, John, ed. "The Complete Work of Charles Darwin Online." http://darwin-online.org.uk/.

Velazquez de la Cadena, Mariano. *A Dictionary of the Spanish and English Languages.* New York: D. Appleton, 1858.

Verkhratsky, Alexei, O. A. Krishtal, and Ole H. Petersen. "From Galvani to Patch Clamp: The Development of Electrophysiology." *Pflügers Archiv—European Journal of Physiology* 453, no. 3 (2006): 233–47.

Vogel, Steven. *Cats' Paws and Catapults: Mechanical Worlds of Nature and People.* New York: W. W. Norton, 1998.

Volmar, Axel. "Listening to the Body Electric: Electrophysiology and the Telephone in the Late 19th Century." http://vlp.mpiwg-berlin.mpg.de/references?id=art76.

Volta, Alexander. "On the Electricity Excited by the Mere Contact of Conducting Substances of Different Kinds." *Philosophical Transactions of the Royal Society of London* 90 (1800): 403–31.

Vrba, E. S. "Ecological and Adaptive Changes Associated with Early Hominid Evolution." In *Ancestors: The Hard Evidence*. Edited by E. Delson. New York: Alan R. Liss, 1985.

Walker, Phillip L. "Butchering and Stone Tool Function." *American Antiquity* 43, no. 4 (1978): 710–15.

Wallace, Alfred Russell. *My Life: A Record of Events and Opinions*. Vol. 1. London: Chapman & Hall, 1905.

———. "On Some Fishes Allied to Gymnotus." *Annals and Magazine of Natural History* 14, no. 83 (1854): 398–99.

Wallace, Wes. "The Vibrating Nerve Impulse in Newton, Willis, and Gassendi: First Steps in a Mechanical Theory of Communication." *Brain and Cognition* 51, no. 1 (2003): 66–94.

Walsh, John. "Of the Electric Property of the Torpedo." *Philosophical Transactions (1683–1775)* 63 (1773): 461–80.

Walter, W. Grey. *The Living Brain*. New York: W. W. Norton, 1953.

Weekes, W. H. "Details of an Experiment in Which Certain Insects, Known as the Acarus Crossi, Appeared Incident to the Long-Continued Operation of a Voltaic Current Upon Silicate of Potass, within a Close Atmosphere over Mercury." *Proceedings of the London Electrical Society* 1, no. 4 (1843): 240–56.

Weeks, John H. "Anthropological Notes on the Bangala of the Upper Congo River." *Journal of the Royal Anthropological Institute of Great Britain and Ireland* 39 (1909): 97–136.

Wehr, Hans. *A Dictionary of Modern Written Arabic*. 3rd ed. Edited by J. Milton Cowan. Ithaca, NY: Spoken Language Services, 1976.

Wengrow, David. "Narmer: Towards the Material Biography of a Name." In *The Archaeology of Early Egypt: Social Transformations in North East-Africa, 10,000 to 2650 BC*. Cambridge: Cambridge University Press, 2006.

Wesley, John. *The Desideratum; or, Electricity Made Plain and Useful by a Lover of Mankind and of Common Sense*. London: Baillière, Tindall & Cox, 1759.

Westby, G. W. Max. "The Ecology, Discharge Diversity and Predatory Behaviour of Gymnotiforme Electric Fish in the Coastal Streams of French Guiana." *Behavioral Ecology and Sociobiology* 22, no. 5 (1988): 341–54.

Westerman, Diedrich H., and Frieda Schütze. *The Kpelle: A Negro Tribe in Liberia*. Göttingen: Vandenhoeck & Ruprecht, 1921.

Whitaker, Harry, C. U. M. Smith, and Stanley Finger, eds. *Brain, Mind, and Medicine: Essays in Eighteenth-Century Neuroscience*. New York: Springer, 2007.

Whitehead, John. *Grammar and Dictionary of the Bobangi Language as Spoken over a Part of the Upper Congo, West Central Africa*. London: Baptist Missionary Society & Kegan, Paul, Trench, Trubner, 1899.

Whiten, Andrew, Kathy Schick, and Nicholas Toth. "The Evolution and Cultural Transmission of Percussive Technology: Integrating Evidence From Palaeoanthropology and Primatology." *Journal of Human Evolution* 57, no. 4 (2009): 420–35.

Whittaker, Victor P. "Arcachon and Cholinergic Transmission." *Journal of Physiology—Paris* 92, no. 2 (1998): 53–57.

———. "The Biochemistry of the Cholinergic Neurone." *Trends in Biochemical Sciences* 1, no. 8 (1976): 172–74.

———. "The Historical Significance of Work with Electric Organs for the Study of Cholinergic Transmission." *Neurochemistry International* 14, no. 3 (1989): 275–87.

Whitteridge, Gweneth. "How Physiology Grew from Anatomy in the Sixteenth and Seventeenth Centuries." *Proceedings of the Physiological Society* (1976): 1P–9P.

———. "Physics and Chemistry Become the Bases of Physiology in the Eighteenth Century." *Proceedings of the Physiological Society* (1976): 9P–16P.

Wieland, Carl. "Ghostly Coincidence in an Unusual Fish." *Creation* 15, no. 4 (1993): 10–11. http://creation.com/ghostly-coincidence-in-an-unusual-fish.

Wiener, Norbert. *Cybernetics, or Control and Communication in the Animal and the Machine*. Cambridge, MA: MIT Press, 1965.

[Wilberforce, Samuel]. "[Review of] *On the Origin of Species*." *Quarterly Review* 108 (1860): 225–64.

Wilkinson, Charles Henry. *Elements of Galvanism, in Theory and Practice*. Vol. 2. London: John Murray, 1804.

Wilkinson, J. G. *Manners and Customs of the Ancient Egyptians*. Vol. 2. London: John Murray, 1837.

Wilkinson, Toby A. H. "What a King Is This: Narmer and the Concept of the Ruler." *Journal of Egyptian Archaeology* 86 (2000): 23–32.

Wilson, George. "On the Electric Fishes as the Earliest Electric Machines Employed by Mankind." *Canadian Journal of Industry, Science, and Art* 3, no. 13 (1858): 58–69.

Wollaston, Arthur N. *English-Persian Dictionary*. London: W. H. Allen, 1882.

Wollaston, William Hyde. "On the Agency of Electricity on Animal Secretions." *Philosophical Magazine* 33 (1809): 488–90.

Wrangham, Richard, Dorothy Cheney, Robert Seyfarth, and Esteban Sarmiento. "Shallow-Water Habitats as Sources of Fallback Foods for Hominins." *American Journal of Physical Anthropology* 140, no. 4 (2009): 630–42.

Wu, C. H. "Electric Fish and the Discovery of Animal Electricity." *American Scientist* 72 (1984): 598–606.

Xu, Jian, and David A. LaVan. "Designing Artificial Cells to Harness the Biological Ion Concentration Gradient." *Nature Nanotechnology* 3 (2008): 666–70.

Yolton, John W. "The Physiology of Thinking and Acting." In *Thinking Matter: Materialism in Eighteenth-Century Britain.* Minneapolis: University of Minnesota Press, 1983.

Zupanc, Günther K. H., and Theodore H. Bullock. "From Electrogenesis to Electroreception: An Overview." In *Electroreception*, edited by Theodore H. Bullock, Carl D. Hopkins, Arthur N. Popper, and Richard R. Fay. New York: Springer, 2005.

# Index

Page numbers followed by *f* indicate figures.